T0185964

Texts in Computer Science

Series Editors

David Gries, Department of Computer Science, Cornell University, Ithaca, NY, USA

Orit Hazzan, Faculty of Education in Technology and Science, Technion—Israel Institute of Technology, Haifa, Israel

More information about this series at http://www.springer.com/series/3191

Wei Qi Yan

Computational Methods for Deep Learning

Theoretic, Practice and Applications

Wei Qi Yan
Auckland University of Technology
Auckland, New Zealand

ISSN 1868-0941   ISSN 1868-095X (electronic)
Texts in Computer Science
ISBN 978-3-030-61083-8   ISBN 978-3-030-61081-4 (eBook)
https://doi.org/10.1007/978-3-030-61081-4

This Springer imprint is published by the registered company Springer Nature Switzerland AG
The registered company address is: Gewerbestrasse 11, 6330 Cham, Switzerland

Preface

This book was drafted based on my recent lectures, talks, and seminars for our postgraduate students at the Auckland University of Technology (AUT), New Zealand. We integrate the materials of deep learning and machine learning as well as artificial neural networks together, refine the content, and publish this book so that more postgraduate students, especially those students who are working for their theses can benefit from our research and teaching work for the purpose of enlightening their projects.

In this book, we organize our stuff and tell our story from easy to difficult in mathematics; we prepare our contents for knowledge transfer from the viewpoint of machine intelligence. We start from understanding artificial neural networks with the design of neurons and the activation functions, then explain the mechanism of deep learning using advanced mathematics. At the end of each chapter, we especially emphasize on how to use Python-based platforms and the latest MATLAB toolboxes for implementing the deep learning algorithms; we also list the questions we concern for thinking and discussion.

Before reading this book, we strongly encourage our readers to learn the knowledge of postgraduate mathematics, especially those fundamental subjects like mathematical analysis, linear algebra, optimizations, computational methods, differential geometry, manifold, information theory as well as basic algebra, functional analysis, graphical models, etc. The computational knowledge will assist us in understanding not only this book but also relevant journal articles and conference papers in the field of deep learning.

This book was written for research students and engineers as well as computer scientists who are interested in computational approaches of deep learning for theoretic analysis and practical development. More generally, this book is also apt for those researchers who are interested in machine intelligence, pattern analysis, computer vision, Natural Language Processing (NLP), and robotics.

Auckland, New Zealand   Wei Qi Yan
September 2020

Acknowledgements

Thanks to our peer colleagues and students whose materials were referenced and who have given invaluable comments on this book. Special thanks to my supervised students: Mr. J. Wang, Dr. Y. Zhang, Mr. J. Lu, Mr. D. Shen, Mr. K. Zheng, Ms. Y. Ren, Mr. R. Li, Mr. P. Li, Mr. Z. Liu, Ms. Y. Shen, Ms. H. Wang, Mr. C. Xin, Ms. Q. Zhang, Mr. C. Liu, Ms. B. Xiao, Ms. X. Liu, Mr. C. Song, Mr. X. Ma, Mr. S. Sun, Ms. Y. Fu, Ms. N. An, Ms. L. Zhang, Dr. Q. Gu, my colleagues Dr. M. Nguyen, Prof. R. Klette.

Contents

About the Author

Wei Qi Yan Auckland University of Technology, Auckland, New Zealand.

Dr. Wei Qi Yan is an Associate Professor with the Auckland University of Technology (AUT), New Zealand; his expertise is in intelligent surveillance, deep learning, computer vision, and multimedia technology; he is the Director of Centre for Robotics & Vision at AUT. He was the Editor-in-Chief (EiC) of the International Journal of Digital Crime and Forensics (IJDCF), now the Editor-in-Chief Emeritus; he was an exchange computer scientist between the Royal Society of New Zealand (RSNZ) and the Chinese Academy of Sciences (CAS), China. He is a guest (adjunct) professor with Ph.D. supervision of the Chinese Academy of Sciences, China; he was a visiting professor of the Massey University, the University of Auckland, New Zealand and the National University of Singapore, Singapore.

Symbols and Acronyms

Symbols

$\mathscr{Z}$	Set of integer numbers
$\mathscr{Z}^+$	Set of positive integer numbers
$\overline{1,n}$	$1, 2, \cdots, n$
$\mathscr{R}$	Set of real numbers
$\cup$	Union of sets
$\cap$	Intersection of sets
$\in$	Member
$\subset$	Proper subset
$\subseteq$	Subset
$\exists$	Exist
$\forall$	For all
$\perp$	Perpendicular
$\triangleq$	Define
$\mapsto$	Mapping
$\pm$	Plus or minus
$\sum$	Sum
$\prod$	Product
∞	Infty
$\|\cdot\|$	Norm function
$det(\cdot)$	Determinant
$\mathbf{N}(\cdot)$	Gaussian distribution
$\sigma(\cdot)$	Activation function
$<\cdot>$	Inner or dot product
$L(\cdot)$	Loss function
$J(\cdot)$	Cost function
$\log(\cdot)$	Logarithm base 10
$\ln(\cdot)$	Natural logarithm
$\exp(\cdot)$	Exponential function
$\tanh(\cdot)$	Hyperbolic tangent function
$\max(\cdot)$	Max function
$\frac{df}{dx}$	Derivative

$\frac{\partial f}{\partial x}$	Partial derivative
$\int$	Integral
C^1	First-order parametric continuity
C^2	Second-order parametric continuity
C^∞	Infinite continuity
$\mathbf{E}(\cdot)$	Expected value function
$p(y\|x)$	Conditional probability
μ	Mean
σ	Variance
$\mathrm{argmax}(\cdot)$	Arguments of the maxima
$sgn(\cdot)$	Sign function
$(w_{ij})_{m\times n}$	Element w_{ij} of $m \times n$ matrix $\mathbf{W}_{m\times n}$
$(b_i)_{n\times 1}$	Element b_i of vector $\mathbf{b}_{n\times 1}$
$\mathbf{W}$	Weight matrix $\mathbf{W}$
$\mathbf{W}^\tau$	Matrix transpose
$\mathbf{b}$	Shift vector $\mathbf{b}$
$\mathbf{b}^\tau$	Vector transpose
P	Point $\mathbf{P}$
S	Set $\mathbf{S}$

Acronyms

ACM	Association for computing machinery
AdaBoost	Adaptive boosting
AI	Artificial intelligence
ANN	Artificial neural networks
ASCII	American standard code for information interchange
Bagging	Bootstrap aggregating
CNN	Convolutional neural network
ConvLSTM	Convolutional long short-term memory
ConvNet	Convolutional neural network
CVPR	International conference on computer vision
DBM	Deep Boltzmann Machine
DBN	Deep belief network
DMRF	Deep Markov random field
FAIR	Facebook AI Research
FCN	Fully convolutional network
FCNN	Fully connected neural network
FN	False negative
FP	False positive
FRU	Fully gated unit
GAN	Generative adversarial network
GPU	Graphics processing unit

GRU	Gated recurrent unit
HOG	Histograms of oriented gradients
ICCV	International conference on computer vision
LBP	Local binary patterns
LSTM	Long short-term memory
MC	Monte Carlo methods
MCNN	Multichannel convolutional neural networks
MDP	Markov decision process
MGU	Minimal gated unit
MNIST	Modified NIST database
MRF	Markov random field
MRI	Magnetic resonance imaging
MRP	Markov random process
NIST	National institute of standards and technology
NLP	Natural language processing
PCA	Principal component analysis
PDF	Probability density function
R-CNN	Region-based CNN
ReLU	Rectified linear unit
ResNet	Deep residual network
RNN	Recurrent neural network
ROI	Region of interest
RPN	Region proposal network
SARSA	State-action-reward-state-action
SGD	Stochastic gradient descent
SSD	Single shot multibox detector
ST-GCN	Spatial-temporal graph convolutional networks
SVM	Support vector machine
TD	Temporal-difference
TN	True negative
TP	True positive
VAE	Variational autoencoder
WCSS	Within-cluster sum of squares
YOLO	You only look once

Introduction

1

1.1 Introduction

Deep learning turns up after many years of evolutions of information technology such as sensor networks, cloud computing, big data, World Wide Web (WWW), mobile technology, supercomputing, etc. Sensor networks provide tremendous data for us to fully touch and understand this cyber world, cloud computing accommodates the storage space for these data. The big data could be visualized and analysed by using WWW and Internet technology, while mobile phones can view or operate the data processing using our thumbs. After this knowledge accumulation and evolution of so many years, deep learning emerges and becomes an iconic technology of this digital era. We can say deep learning is a subsequence of historical necessity.

Deep learning is a redhot technology at present, which is thought as a core technology in Artificial Neural Networks (ANNs) and Artificial Intelligence (AI). Deep learning is also named as deep neural networks (DNNs), ANNs is the core content of AI. The latest development of AI has been reflected in deep learning.

ACM 2018 Turing Award has been conferred to a trio of computer scientists: Yoshua Bengio, Geoffrey Hinton, and Yann LeCun for their conceptual and engineering breakthroughs that have made DNNs a critical component of computing in 2019. Turing Award, which usually is thought as the Nobel Prize of Computing, is an annual prize given by the Association for Computing Machinery (ACM) to the persons selected for contributions of lasting and major technical importance to the computer field.

The articles published by this group of computer scientists in the journals Nature [1] and Science [2] have shown their distinctive contributions to this field. The publications have been regarded as the classical work of this field. The book entitled deep learning published by the MIT press [3] outlined this research area and inspired many young students and enthusiasts. As stated in the book, pertaining to DNNs, usually the input data is imported and the outputs from those activation functions for simulating the stimuli of neurons are calculated, the activation functions include ReLU function, sigmoid function, logistic function, etc. Transfer function comes

© The Author(s), under exclusive license to Springer Nature Switzerland AG 2021
W. Q. Yan, *Computational Methods for Deep Learning*, Texts in Computer Science,
https://doi.org/10.1007/978-3-030-61081-4_1

from the name transformation which is used for transformation purposes, i.e., from input nodes to the output of a neuron. On the other hand, activation function checks for the output if it meets a threshold and either outputs zero or one. The difference between input data and output data of a neural network should be minimized.

In 2011, ReLU activation function $y = x^+ = \max(x, 0), x \in (-\infty, +\infty)$ was found much better than Tanh activation function $y = \tanh(x), x \in (-\infty, +\infty)$ in resolving the vanishing gradient problem, which pave the ways for further development of deeper neural networks [4].

Usually, we use backpropagation including forward pass and backward pass to adjust the weights of convolutional neural works. The algorithms of DNNs from machine learning have been grouped into supervised learning and unsupervised learning. Supervised learning is related to labelled data or ground truth. Public websites such as NIST (National Institute of Standards and Technology) also provide verified dataset MNIST (Modified NIST database) for training and testing deep learning models. On the contrary, unsupervised learning is subject to the similarity functions, for instance, clustering is a typical unsupervised learning. In deep learning, the unsupervised learning methods include Principal Component Analysis (PCA), autoencoder, manifold learning, etc. The unsupervised learning approaches have been applied to dimensionality reduction. Dimensionality reduction, or dimension reduction, is the transformation of data from a high-dimensional space into a low-dimensional space so that the low-dimensional representation retains meaningful properties of the original data, ideally close to its intrinsic dimension.

The famous playground software Tinker has been applied to understand ANNs as shown in Fig. 1.1 which renders how the DNNs work, the relevant parameters and outcomes are explicitly linked to one web page. Four types of examplar datasets are provided for demonstrations. Any adjustments of the input parameters will reflect the changes of the visualized results of classification and regression. The network

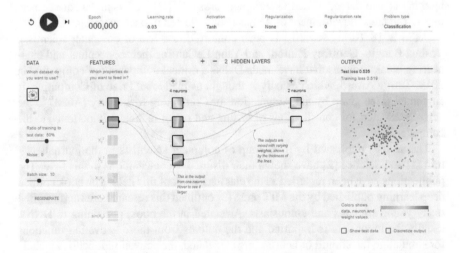

Fig. 1.1 The playground software for understanding neural networks

architecture could be manually adjusted, the nodes of the network could be freely added or removed by a network designer. L_1 and L_2 regularizations have been provided for selection. The other options include four activation functions, learning rates, and epoch numbers.

In the course of backpropagation, we need to calculate the stochastic gradient based on optimization, usually SGD (Stochastic Gradient Descent) will be adopted. SGD is an iterative method for optimizing an objective function with suitable smoothness properties (e.g., differentiable or subdifferentiable). Thus, the chain rule for function differentials is required. Meanwhile, minibatch has been taken into account for various parametric optimization.

In deep learning, the hottest research topics now are distributed in the fields of manifold learning, reinforcement learning, Generative Adversarial Network (GAN), etc. These typical approaches have been applied to automation or automatic control, robotics, machine vision, natural language processing, intelligent surveillance, recommender systems, etc.

Deep learning is updated very quickly. The site Github.com provides relevant source codes and datasets for various projects. From a technological viewpoint, there are two very popular software platforms: MATLAB and Python-based TensorFlow which could be applied to implement deep learning projects easily.

1.2 Deep Learning

Deep learning, also called deep neural networks, is originated from modelling biological vision and brain information processing. Deep learning is one part of the contents of machine learning or machine intelligence. AlexNet has taken the first step of deep learning, which has been applied to handwriting recognition with the famous MNIST dataset, i.e., AlexNet was designed by Alex Krizhevsky and published with Ilya Sutskever and Geoffrey Hinton. AlexNet was competed in the ImageNet Large Scale Visual Recognition Challenge on September 2012 [5]. The network achieved a top 5 error of 15.3%, lower than that of the runner up. In 2015, AlexNet won the ImageNet 2015 contest [6, 7]. ImageNet is an image database organized according to the WordNet hierarchy, in which each node of the hierarchy is depicted by hundreds and thousands of images.

Deep learning is closely related to mathematics, especially optimization, graphical theory, numerical analysis, functional analysis, probability theory, mathematical statistics, information theory, etc. These subjects could provide the analysis for a neural network model. Usually, when we measure a neural network or evaluate an algorithm, we take into consideration of its robustness, stability, and convergence in numerical analysis.

In deep learning, we use gradient descent to update the parameters of our deep learning models. Gradient descent is a first-order iterative optimization algorithm for finding the local minimum of a function. For example, gradient descent is used to solve a system of linear equations as a quadratic minimization problem, e.g., using linear least squares. The solution of

$$\mathbf{A}\mathbf{x} - \mathbf{b} = 0 \tag{1.1}$$

is defined as minimizing the function

$$F(\mathbf{x}) = \|\mathbf{A}\mathbf{x} - \mathbf{b}\|_2^2. \tag{1.2}$$

In linear least squares for real $\mathbf{A}$ and $\mathbf{b}$, the Euclidean norm is used,

$$\nabla F(\mathbf{x}) = 2\mathbf{A}^T(\mathbf{A}\mathbf{x} - \mathbf{b}). \tag{1.3}$$

With this observation in mind, one starts with a guess $\mathbf{x}_0$ for a local minimum of F, and considers the sequence $\mathbf{x}_0, \mathbf{x}_1, \mathbf{x}_2, \ldots$ such that

$$\mathbf{x}_{n+1} = \mathbf{x}_n - \gamma \nabla F(\mathbf{x}_n), \ n \geq 0, \tag{1.4}$$

where $\gamma \in \mathscr{R}^+$ is small enough, the value of step size γ is allowed to be adjusted at each iteration.

If we have a cost or error function $F(\mathbf{w})$ that needs to be minimized, the gradient descent tells us to modify the weights in the direction of the steepest descent in $F(\mathbf{w})$, the weight decay equation is

$$\mathbf{w}_{n+1} = \mathbf{w}_n - \gamma \nabla F(\mathbf{w}_n), \ n \geq 0, \tag{1.5}$$

where γ is the learning rate, $\mathbf{w}_n$ represents the weights of DNNs. In numberical analysis [8], the terminal conditions for the iterations in Eq. (1.5) are a preset number of loops or running time as well as a given resultant estimation and the errors between two adjacent loops.

In deep learning, we are still facing the optimization problems such as solution existence, stability, robustness, and convergence for weight decay like most of the existing ones in computational methods. The two problems in deep learning are gradient vanishing problem and gradient exploding problem.

At present, gradient vanishing problem and gradient exploding problem in SGD could be resolved by making use of multilevel hierarchy of networks restricted Boltzmann machine, generative model, long short-term memory (LSTM), residual networks (ResNets) while gradient exploding could be avoided by using redesigned network, ReLU activation functions, LSTM in RNN, gradient clipping, weight regularization, etc.

Regularization is applied to avoid gradient exploding and vanishing [3, 9]. Regularization is defined as a modification we make to a learning algorithm that is to reduce its generalization errors but not its training errors. Regularization will help to reduce overfitting and drive the weights to lower values.

The regularized objective function is

$$\hat{J}(\theta; \mathbf{X}, \mathbf{y}) = J(\theta; \mathbf{X}, \mathbf{y}) + \alpha \cdot \Omega(\theta), \tag{1.6}$$

where $\alpha \in [0, \infty)$ is a hyperparameter or regularization rate; θ denotes all of the parameters. The optimized parameter θ^* is obtained by using

$$\theta^* = \arg\min_{\theta} \nabla_{\theta} \hat{J}(\theta; \mathbf{X}, \mathbf{y}), \tag{1.7}$$

The typical regularizations include Tikhonov regularization, sparse regularization, Lagrangian regularization, etc.

Tikhonov Regularization or L_2 regularization is

$$\Omega(\theta) = \frac{1}{2}\|\mathbf{w}\|_2^2. \tag{1.8}$$

Thus,

$$\hat{J}(\mathbf{w}; \mathbf{X}, \mathbf{y}) = J(\mathbf{w}; \mathbf{X}, \mathbf{y}) + \frac{\alpha}{2}\mathbf{w}^{\tau}\mathbf{w}. \tag{1.9}$$

The gradient,

$$\nabla_{\mathbf{w}} \hat{J}(\mathbf{w}; \mathbf{X}, \mathbf{y}) = \nabla_{\mathbf{w}} J(\mathbf{w}; \mathbf{X}, \mathbf{y}) + \alpha \cdot \mathbf{w}. \tag{1.10}$$

To update the weights,

$$\mathbf{w} \leftarrow \mathbf{w} - \epsilon \cdot \nabla_{\mathbf{w}} \hat{J}(\mathbf{w}; \mathbf{X}, \mathbf{y}), \epsilon \in (0, 1). \tag{1.11}$$

Sparse regularization or L_1 regularization refers to

$$\Omega(\theta) = \|\mathbf{w}\|_1 = \sum_i |\omega_i| \tag{1.12}$$

L_1 regularization is

$$\hat{\mathbf{J}}(\mathbf{w}; \mathbf{X}, \mathbf{y}) = \alpha \|\mathbf{w}\|_1 + J(\mathbf{w}; \mathbf{X}, \mathbf{y}). \tag{1.13}$$

With regard to the gradient,

$$\nabla_{\omega} \hat{\mathbf{J}}(\mathbf{w}; \mathbf{X}, \mathbf{y}) = \alpha \cdot Sign(\mathbf{w}) + \nabla_{\omega} J(\mathbf{w}; \mathbf{X}, \mathbf{y}) \tag{1.14}$$

$$\mathbf{w} \leftarrow \mathbf{w} - \epsilon \cdot \nabla_{\mathbf{w}} \hat{J}(\mathbf{w}; \mathbf{X}, \mathbf{y}), \epsilon \in (0, 1). \tag{1.15}$$

Lagrangian regularization means a generalized Lagrange function (or multiplier) and a constant k satisfy

$$\mathcal{L}(\theta, \alpha; X, y) = J(\theta; X, y) + \alpha \cdot (\Omega(\theta) - k). \tag{1.16}$$

The solution is

$$\theta^* = \arg\min_{\theta} \mathcal{L}(\theta, \alpha). \tag{1.17}$$

If we have a fixed α^*, then

$$\theta^* = \arg\min_{\theta} \mathcal{L}(\theta, \alpha^*) = \arg\min_{\theta} J(\theta; X, y) + \alpha^* \cdot \arg\min_{\theta}(\Omega(\theta) - k). \tag{1.18}$$

When we seek the maxima in weight decay, we cannot guarantee all the weights exist that could be found directly. But after regularization, we can better obtain the peak point and avoid the tough problems if we use stochastic gradient descent (SGD).

The mathematical regularization has plenty of merits, by using increased bias for reduced variance, regularization can reduce overfitting and drive the weights to lower values. The regularization is as effective as a dropout.

1.3 The Chronicle of Deep Learning

Deep learning has shown its effectiveness and superiority for resolving practical problems. It has been particularly successful in surpassing our human capabilities in visual object detection and recognition, image segmentation, speech recognition, natural language processing, and robot control when the technology related to big data is employed.

In 1995, CNN (convolutional neural network, or ConvNets) as a typical neural work has been employed for postcode recognition [10–12] on an envolop or handwritten numbers on a bank check. CNNs assist us to find ROI (i.e., region of interest) and salient regions, which are based on emulating the mechanism of our human neural system. The end-to-end structure and fine-tuning merits inspire us recognizing tiny objects with details at pixel level [13]. Typically, LeNet-5, a pioneering 7-level convolutional network created by LeCun Yann, et al. in 1998 [14] that classifies digits, was applied by several banks to recognize handwritten numbers on bank checks digitized in 32×32 pixel images.

In 1997, AdaBoost algorithm was proposed for boosting a strong classifier from a weak one [15, 16]. This makes ensemble learning could be applied to machine learning [17] or deep learning [3].

Random forests are an ensemble learning method for classification and regression by constructing a multitude of decision trees at training time and outputting the class that is the mode of classification regression of individual trees. Random forest [18] is based on decision trees, a random forest is shown as Fig. 1.2. A random forest [18] is formed when multiple trees are ensembled together [19]. Decision trees usually are employed for decision-making [17].

Since 1995, SVM (support vector machine) [20] has become a popular machine learning algorithm for pattern classification based on specific features as well as the hyperplane and hyperparameters [17]. Different from SVM [20] in machine learning, deep learning algorithms use multiple classes, the class with the highest probability will be thought as the output of a neural network [21]. Deep learning [3] is based on the end-to-end framework of neural networks, which is a well-designed subject not only in programming and data collection but also in mathematical theories and network structures.

Deep Belief Network (DBN) is a directed network [22, 23], where the edges and nodes have different weights; on the other hand, deep Markov random field [24] is an undirected network, where all the edges are bidirectional. A Deep Boltzmann

Machine (DBM) is a type of binary pairwise Markov random field (undirected probabilistic graphical model) with multiple layers of hidden random variables. It is a network of symmetrically coupled stochastic binary units [25, 26]. DBM has been successfully applied to pattern classification, regression, and time series modelling.

AlexNet is an early simple neural network [6]. AlexNet contains eight layers, the first five were convolutional layers, followed by max pooling layers, the last three were fully connected layers. As a milestone of machine learning and machine intelligence, AlexNet won the ImageNet challenge 2012. In its further version, deep learning surpasses our human vision system in the test pertaining to computer vision. AlexNet was implemented by using MATLAB in its early deep learning toolbox and has been developed for transfer learning.

AlexNet is a CNN network that is trained based on more than a million of images from the ImageNet database. The network is eight layers depth and can classify images into 1,000 object categories, including keyboards, mouses, pencils, and many animals. The network has an image input size of 227×227.

R-CNN is the region-based CNN. Based on traditional CNNs, the proposals have been added into this structure of neural networks. The proposed region of interest has been recommended to reduce the processing time.

Fast [27–29] and Faster R-CNN [30, 31] are region-based CNN (R-CNN) which was originated from CNN (ConvNets), but R-CNN has been applied to quickly find the object based on region segmentation and ROI [30]. If the region could be given much earlier, that will speed up the computations of visual object location and classification.

The biggest problem of R-CNN is that its training time is very long because it needs to get 1,000–2,000 proposals first and save them; these proposals need to be calculated in all the former layers. In addition, the fully connected layer is expected that all the vectors will have the same size, all the proposals need to be resized

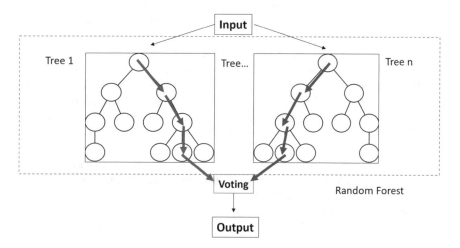

Fig. 1.2 A random forest

using cropping or wrapping, both strategies are not suitable because the cropping may cause the proposals are not fully extracted and the wrap could change scales of objects.

Fast R-CNN was proposed in 2015 which overcomes several problems of R-CNN. What Fast R-CNN has done is to replace ROI pooling layer, and softmax is applied to the classification. The softmax is one extension of logistic regression function to the multiclassification problem.

After Fast R-CNN, Faster R-CNN was proposed to improve the training speed of the Fast R-CNN. From R-CNN to Faster R-CNN, four steps of object detection are finally unified into one network. Faster R-CNN does not use selective search to get region proposals. Instead, it makes use of a region proposal to carry out the same task. There is no repetition and all the calculations are performed by using GPUs [30, 31].

In recent years, YOLO [32] (Darknet) has become a very popluar deep network. Darknet is an open-source framework. It is fast, easy to be installed, which supports CPU and GPU computations. YOLOv3 is the third version of YOLO family (https:// pjreddi.e.com/darknet/). Before YOLOv3, YOLO, and YOLOv2 already have been developed for visual object detection in deep learning, especially for pedestrian detection [33]. In 2020, YOLOv4 has been proposed for optimal speed and accuracy of object detection.

YOLO is one of the fast object detectors, which creates grid cells, each cell will predict the bounding box and the confidence score of this box. For evaluating YOLO, a 7×7 bounding box and 20 labelled classes are defined, which means, it only extracts 98 proposals. YOLO is faster than R-CNN which needs 2,000 proposals.

YOLO uses the whole image instead of a region proposal to train and test which has a lower rate of background error. When compared to another real-time system on PASCAL VOC 2007, YOLO has an overwhelming advantage. Fast R-CNN takes around 2 s per image to generate bounding box proposals. Faster R-CNN as the most accurate model reaches 7 fps while a smaller model, which has lower mPA of 62.1 (Faster R-CNN), achieves 18 fps. But YOLO could reach 45 fps, which is twice faster than R-CNN and even has a higher mPA of 63.4. The limitation of YOLO is that each cell could predict bounding boxes but there is only one class that could be detected, which makes small objects hardly to be detected.

SSD (single shot multibox detector) [34, 35] is famous for balancing the resolution and speed as shown in Fig. 1.3. Meanwhile, YOLO [32] and YOLOv2 are excellent in achieving the fast speed to detect objects based on operations using 7×7 blocks. Now, YOLOv3 has been thought to overcome the shortcomings and become a very excellent method for object detection.

Deep residual network (ResNet) [36, 37] was designed for avoiding the problems of vanishing gradients and exploding gradients, reusing activations from a previous layer until the adjacent layer learns its weights. ResNets utilize skip connections, or shortcuts to jump over some layers.

Fig. 1.3 SSD for visual object recognition

In deep learning, there exists the degradation problem, namely, with the network depth increasing, accuracy gets saturated. However, ResNets can easily enjoy accuracy gains from greatly increased depth.

$$\mathbf{y} = \mathscr{F}(\mathbf{x}, \{W_i\}) + \mathbf{x}, \tag{1.19}$$

where $\mathbf{x}$ and $\mathbf{y}$ are the input and output vectors of the layers, $\mathscr{F}(\cdot)$ is the residual mapping, e.g., $\mathscr{F} = W_{2\sigma}(W_{1x})$, $\sigma(\cdot)$ is the ReLU activation function.

GAN refers to generative adversarial network [38, 39] that could be applied to identify the differences between a fake object and the real one. GAN is a type of deep learning network that can generate data with similar characteristics as the input training data. A GAN consists of two networks that train together: *generator* and *discriminator*. In order to train a GAN, it needs to train the generator to generate data that fools the discriminator and train the discriminator to distinguish between real and generated data. The objective of the generator is to generate data that the discriminator classifies as the real one. The objective of the discriminator is not to be fooled by the generator. These strategies result in a generator that generates convincingly realistic data and a discriminator that has learned strong feature representations.

Reinforcement learning [40, 41] is suitable for pendulum control. An environment and relevant states, as well as rewards, are established for an agent to control the system. Once the agent takes a step, we need to calculate the rewards and assess the feedback so as to decide whether the action is with positive or negative reward [17].

In Markov decision processes, a Bellman equation is a recursion for expected rewards

$$V^\pi(s) = R(s, \pi(s)) + \gamma \sum_{s'} P(s'|s, \pi(s)) V^\pi(s'), \qquad (1.20)$$

where s is a state and the policy is π, $V(\cdot)$ is the value function, $R(\cdot)$ is the reward function. The Bellman optimality equation is

$$V^{\pi^*}(s) = \max_a \{R(s, a) + \gamma \sum_{s'} P(s'|s, a) V^{\pi^*}(s')\}, \qquad (1.21)$$

where π^* is the optimal policy and V^{π^*} refers to the value function of the optimal policy.

In recent years, Q-learning algorithm is proposed for solving the problem given by Bellman equation. The goal of Q-learning algorithm [42] is to learn a policy, which tells an AI agent what action will take under what circumstances.

An autoencoder is used to learn efficient data codings in an unsupervised manner [43]. The aim of an autoencoder is to learn a representation (encoding) for a set of data, typically for dimensionality reduction, by training the network to ignore signal noise [44].

Given one hidden layer, the encoder stage of an autoencoder takes the input $\mathbf{x} \in \mathbf{R}^d = \mathbf{X}$ and maps it to $\mathbf{h} \in \mathbf{R}^p = \mathbf{F}$

$$\mathbf{h} = \sigma(\mathbf{W}\mathbf{x} + \mathbf{b}), \qquad (1.22)$$

where $\mathbf{h}$ is usually referred to latent variables, $\sigma(\cdot)$ is an activation function, $\mathbf{W}$ is a weight matrix, and $\mathbf{b}$ is a bias vector.

Autoencoders are trained to minimize reconstruction errors as the loss

$$\mathcal{L}(\mathbf{x}, \mathbf{x}') = \|\mathbf{x} - \mathbf{x}'\|^2 = \|\mathbf{x} - \sigma'(\mathbf{W}'(\sigma(\mathbf{W}\mathbf{x} + \mathbf{b})) + \mathbf{b}')\|^2, \qquad (1.23)$$

where $\mathbf{x}$ is usually averaged over the input training set. Mathematically, this loss function is a square loss function. A family of loss functions also include 0–1 loss function, absolute loss function, average loss function, etc.

Autoencoders are assisted by using the output as its input iteratively, which leads to the famous fixed-point theorem. If the fixed point could be found or converged, that means we can use autoencoders for noises removal and dimension reduction. Autoencoders have been successfully applied to image inpainting and image denoising [45].

Transfer learning [46] is to apply the well-trained weights of an existing neural network to a new model. We modify the weights or parameters and make minor corrections. It saves training time, however, does not affect the average precision of patter classification [47] too much.

Transfer learning applies stored knowledge to a different but related problem. From a practical standpoint, reusing or transferring information from previously learned tasks for the learning of new tasks has the potential to significantly improve the sample efficiency in deep learning.

In the new version of MATLAB, there is a toolbox especially for implementing transfer learning [5]. In transfer learning, a deep learning approach in which a model that has been trained for one task is used as a starting point to train a model for similar task. Fine-tuning a network with transfer learning is usually much faster and easier than training a network from scratch. Transfer learning is usually used for visual object detection, image recognition, speech recognition, and other applications.

Markov random fields (MRFs) [48] are undirected network which relates to the adjacent state. We assume the Markov random process [49] only has effects in the next timeframe in time series analysis.

In graphical models [49], MRF and DBN are two typical networks. Dynamic Bayesian networks, such as influence diagrams [50], are thought as the directed network, the probability of each node is completely dependent on its neighbours. Therefore, conditional probability and joint probability are especially required.

SqueezeNets [51] and compressed networks(CompressedNets) are used for mobile phones or small devices. If we trim a network or compress the network, we could use it for the small devices which has not very big memory or needs GPU to support the computations though some tablets and mobiles now have facilitated with these hardware.

Ensemble learning [52, 53] integrates all learners together. By using ensemble learning, we boost a weak classifier to strong. The typical ones are AdaBoost and Bagging algorithms. We have employed this algorithm in OpenCV for visual object detection.

Entropy is a measure of the unpredictability of the state, or equivalently, of its average information. The measure of information entropy associated with each possible data value is the negative logarithm of the probability mass function for the value

$$S = -\sum_i P_i \log P_i = -\mathbf{E}_P(\log P). \tag{1.24}$$

where $\mathbf{E}_P(X) = \sum_i P_i X_i$ is mathematical expectation defined by the probability P. Function $\mathbf{E}_P[X]$ has the properties:

- $\mathbf{E}_P(c) = c$, c is a constant.
- $\mathbf{E}_P(cX) = c\mathbf{E}_P(X)$, c is a constant.
- $\mathbf{E}_P[\mathbf{E}_P(X)] = \mathbf{E}_P(X)$.
- $\mathbf{E}_P(X \pm Y) = \mathbf{E}_P(X) \pm \mathbf{E}_P(Y)$.
- $\mathbf{E}_P(aX \pm b) = a\mathbf{E}_P(X) \pm b, a, b \in R$.
- If X and Y are independent, $\mathbf{E}_P(XY) = \mathbf{E}_P(X)\mathbf{E}_P(Y)$.

Entropy [54, 55] is a fundamental concept in information theory [56]. All kinds of entropy, such as joint entropy, conditional entropy, mutual information (KL divergence) construct the main body of information theory. Information theory [56] has been applied to information retrieval [57, 58] previously, now it has been employed to deep learning [3].

Mathematics refers to calculus [59] and algebra [60]. Calculus is related to numerical analysis; linear algebra, and tensor algebra are employed in deep learning. Tensor algebra [60] is needed for better understanding TensorFlow [61] from Deepmind of Google Inc.

Besides these, in deep learning, we need to study probability theory and mathematical statistics, we also need to understand functional analysis [62] and abstract algebra [63] for fully understanding the algorithms and mechanism for the cost functions in a normed space and polynomial rings.

Since we have known deep learning was not fallen from the sky, it accumulates human computing experience in the past decades from sensor networks, big data, cloud computing, image processing, computer vision, pattern classification, and machine learning, etc.

We strongly suggest a beginner should start the deep learning study from a well-written paper [1, 64, 65] or a good book [3], download the relevant source codes for implementation and experiencing the differences of deep learning and general machine learning. Based on these first-hand experience, further deep learning study or research work is encouraged. Mathematical knowledge and network structure are especially recommended.

1.4 Our Deep Learning Projects

A myriad of projects have been developed based on deep learning [66–74]. The features of these development are different from traditional machine learning or pattern classification. We would like to list some of them below.

A project has been developed for human face detection and recognition [70, 75, 76]. In this project, we took more face photos from multiple views and train the inception network using data augmentation. If a face could not be detected, we will quickly switch it to human gait recognition. If a face has been partially covered, we can still detect it using the well-trained model and training dataset.

In human age estimation based on face recognition [77], we propose an improved end-to-end learning algorithm to address the aggregation of multiclass classification and regression for age estimation by using deep CNNs. Our contribution is an updated algorithm for age estimation by adopting the latest attention and normalization mechanisms for balancing the efficiency and accuracy of the proposed model. In addition, our model suits to be deployed on mobile owing to its compact size and superior performance. In future, we will explore this mechanism for other applications related to facial information.

Human behaviour such as running, jumping, skipping, etc., could be detected using deep learning [78]. Based on this work, we quickly detect pedestrians, especially abnormal behaviours for anomaly detection [79]. The differences from previous work based on Local Binary Patterns (LBP) and Histograms of Oriented Gradients (HOG) are that we applied a big training dataset with YOLOv3 as the classifier.

A novel model of dynamic skeletons called Spatial–Temporal Graph Convolutional Networks (ST-GCN) was proposed by automatically learning both the spatial and temporal patterns from visual data to achieve the human behaviour recognition. It performs pose estimation based on videos and constructs spatiotemporal graph of skeleton sequences. Multilayers of spatiotemporal graph convolution gradually generate high-level feature maps on the graph. It is classified by applying the standard softmax classifier to the corresponding category.

Human gait recognition is one of the most promising biometric technologies, especially for unobtrusive video surveillance and human identification from a distance [76, 80–83]. Aiming at improving the recognition rate, we study gait recognition using deep learning and proposed a method based on multichannel convolutional neural networks(MCNN) and convolutional long short-term memory (ConvLSTM).

Human finger motion could be detected for Morse code enter [84, 85]. In a circumstance, if we are not allowed to speak loudly, we can use gesture or Morse codes to contact others. Writing Morse codes on a table surface could not attract much attention. Computer vision techniques using human gesture recognition could help in mutual silent communications in mute mode.

In recent years, DNNs have achieved a remarkable progression in resolving complicated problems. DNNs are suitable for dealing with the problems related to time series analysis, such as speech recognition and natural language processing. Video dynamics detection, as an instance, is time dependent. Apparently, video dynamics detection needs to utilize the present, previous and next frames of a given video. If a frame change occurs, it triggers whether a video event happens or not. We are able to achieve high-precision and real-time video dynamics detection by using RNN and GRU as well as apply CNN to reduce video size and extract critical information. We integrate CNN and RNN together to make a significant reduction in the size of video data and the training time [86].

Blindspot detection [87] of moving vehicles has been developed as a research project. The frequently look back action of a driver could be reduced, a monitoring system could automatically and timely count moving objects in the blindspots and report potential hazards to the driver in all time.

Flame detection [72, 88] is a series of projects we have developed for many years; the flames from burning fires and combustible materials could be recognized. In this project, we detect the flame region with fine-tuning from deep learning. The features of natural fire have been found to discriminate the right or wrong flames.

Currency and coin detection, recognition, and forensics are still very important [89, 90]. So far, we can quickly find the currency from the aspect of object detection. We mainly use SSD model based on deep learning as the framework, employ CNN to extract the features of paper currency, so that we can accurately recognize the denomination of the currency and coin, both front and back.

We used deep learning methods for removing the noises from an image. After training a neural network model, we can obtain parameters to make any picture smooth including the picture after JPEG lossy compression. Deep learning has the capability to reconstruct an image.

Alzheimer's disease (AD) is a neurodegenerative disorder which leads to memory and behaviour impairment [91–93]. Early discovery and diagnosis can delay the progress of this disease. Deep learning has been applied to Alzheimer's disease diagnosis. Selective kernel network with attention(SKANet) for early diagnosis of AD using magnetic resonance imaging (MRI) is proposed. Attention mechanisms have become an integral part of compelling sequence modelling and transduction models in various tasks, allowing modelling of dependencies with regard to the distance in the input or output sequences. The attention mechanism is added to the bottom of the block to emphasize on important features and suppress unnecessary ones for an accurate representation of the network [94].

Deep learning can aid fruit recognition and allow a computer to detect a fruit and find its freshness and ripeness automatically [95]. Apple ripeness identification is such a type of pattern classification. In this project, the ripeness of apples will be detected with DNN and CNN will be employed. The goal is to verify the capability of deep learning in apple recognition for orchards to reduce workers' human labour. Furthermore, deep learning could be applied to food safety assessment, we have employed deep learning to meat quality analysis [73, 74, 96, 97].

1.5 Awarded Work in Deep Learning

In this section, we mainly emphasize the awarded work on IEEE CVPR (International conference on computer vision and pattern recognition) and IEEE ICCV (International conference on computer vision) conferences. IEEE ICCV (1987–present) has the best paper award—Marr prize. The prize is regarded as one of the top honours for a computer vision researcher.

In 2019, the paper "SinGAN: Learning a generative model from a single natural image" has been selected and awarded in ICCV'19, generative deep learning models have been deeply investigated and appeared in most of pattern recognition conferences and workshops of that year.

In 2017, the paper "Mask R-CNN from Facebook AI Research (FAIR)" has been selected. Mask R-CNN is simple to be trained which adds only a small overhead to Faster R-CNN [31]. Mask R-CNN outperforms all existing, single-model entries on every task.

In 2015, the award was conferred to the paper "Deep Neural Decision Forests" [18] from Microsoft Research Cambridge (UK) which is an approach that unifies classification trees with the representation learning functionality known from deep convolutional networks, by training them in an end-to-end manner.

IEEE CVPR (1985–present) conference is regarded as the biggest academic conference of this world, which is highly selective with generally less 30% acceptance rates for all papers and less 5% for oral presentations. The conference is usually held in June and rotates around the US generally West, Central, and East.

A good paper is reflected in many aspects such as its idea, writing, references, equations, tables, and figures, etc. The key is how the paper attacks readers and what the impact could be generated from this work. In recent years, a great deal of deep learning papers has been awarded in the CVPR conferences.

In 2018, the best paper of CVPR was "Taskonomy: Disentangling Task Transfer Learning" [46]. Transfer learning has been published because of the hot research in deep learning.

In 2017, the paper entitled "Densely Connected Convolutional Networks" has been selected and awarded. DensNet [98] was regarded as the most important work of that year.

In 2016, the paper "Deep Residual Learning for Image Recognition from Microsoft Research" has been awarded. ResNet [53] was thought as a prominent contribution of that year.

Meanwhile, the famous academic journals Nature and Science have published multiple papers related to deep learning [1, 41, 99–101] and [2, 102, 103]. These publications have ramped up the research and led the deep learning research work to much deeper.

1.6 Questions

Question 1. Where does deep learning come from?

Question 2. Why deep learning is so important in AI study?

Question 3. What are gradient vanishing and gradient exploding in deep learning? How to avoid them?

Question 4. What is the difference between deep learning methods and SVM (support vector machine)?

Question 5. Why does deep learning affect computer vision, image and video technology so much?

References

1. LeCun Y, Bengio Y, Hinton G (2015) Deep learning. Nature 521:436–444
2. Hinton GE, Salakhutdinov RR (2006) Reducing the dimensionality of data with neural networks. Science 313(5786):504–507
3. Goodfellow I, Bengio Y, Courville A (2016) Deep learning. MIT Press, Cambridge
4. Glorot X, Bordes A, Bengio Y (2011) Deep sparse rectifier neural networks. In: International conference on artificial intelligence and statistics, pp 315–323
5. Krizhevsky A, Sutskever I, Hinton GE (2012) ImageNet classification with deep convolutional neural networks. In: Advances in neural information processing systems, pp 1097–1105
6. Krizhevsky A, Sutskever I, Hinton G (2017) ImageNet classification with deep convolutional neural networks. Commun ACM 60(6):84–90
7. Kriegeskorte N (2015) Deep neural networks: a new framework for modelling biological vision and brain information processing. Ann Rev Vis Sci 24:417–446

8. Stoer J, Bulirsch R (1991) Introduction to numerical analysis, 2nd edn. Springer, Berlin
9. Wan L, Zeiler M, Zhang S, Le Cun Y, Fergus R (2013) Regularization of neural networks using DropConnect. In: International Conference on Machine Learning, pp 1058–1066
10. LeCun Y, Boser B, Denker JS, Henderson D, Howard RE, Hubbard W, Jackel LD (1989) Backpropagation applied to handwritten zip code recognition. Neural Comput 1(4):541–551
11. Tang A, Lu K, Wang Y, Huang J, Li H (2015) A real-time hand posture recognition system using deep neural networks. ACM Trans Intell Syst Technol (TIST) 6(2):21
12. LeCun Y, Bengio Y (1995) Convolutional networks for images, speech, and time series. In: The Handbook of Brain Theory and Neural Networks, vol 3361, issues 10. MIT Press, Cambridge
13. Lee CY, Gallagher PW, Tu Z (2016) Generalizing pooling functions in convolutional neural networks: mixed, gated, and tree. In: Artificial Intelligence and Statistics, pp 464–472
14. LeCun Y, Bottou L, Bengio Y, Haffner P (1998) Gradient-based learning applied to document recognition. Proc IEEE 86(11):2278–2324
15. Ertel W (2017) Introduction to artificial intelligence. Springer International Publishing, Berlin
16. Norvig P, Russell S (2016) Artificial intelligence: a modern approach. 3rd edn. Prentice Hall, Upper Saddle River
17. Alpaydin E (2009) Introduction to machine learning. MIT Press, Cambridge
18. Kontschieder P, et al (2015) Deep neural decision forests. ICCV
19. Gottschalk S, Lin MC, Manocha D (1996) OBBTree: a hierarchical structure for rapid interference detection. In: Conference on computer graphics and interactive techniques, pp 171–180
20. Yeh CY, Su WP, Lee SJ (2011) Employing multiple-kernel support vector machines for counterfeit banknote recognition. Appl Soft Comput 11(1):1439–1447
21. Zanaty EA (2012) Support vector machines (SVMs) versus multilayer perception (MLP) in data classification. Egypt Inf J 13(3):177–183
22. Hinton GE, Osindero S, Teh YW (2006) A fast learning algorithm for deep belief nets. Neural Comput 18(7):1527–1554
23. Sarikaya R, Hinton GE, Deoras A (2014) Application of deep belief networks for natural language understanding. IEEE/ACM Trans Audio Speech Lang Process 22(4):778–784
24. Blake A, Rother C, Brown M, Perez P, Torr P (2004) Interactive image segmentation using an adaptive GMMRF model. In: European conference on computer vision, pp 428–441. Springer, Berlin
25. Fischer A, Igel C (2012) An introduction to restricted Boltzmann machines. In: Iberoamerican congress on pattern recognition, pp 14–36
26. Ackley DH, Hinton GE, Sejnowski TJ (1987) A learning algorithm for Boltzmann machines. In: Readings in computer vision, pp 522–533
27. Girshick R, Donahue J, Darrell T, Malik J (2016) Region-based convolutional networks for accurate object detection and segmentation. IEEE Trans Pattern Anal Mach Intell 38(1):142–158
28. Girshick R (2015) Fast R-CNN. In: IEEE international conference on computer vision, pp 1440–1448
29. Gkioxari G, Girshick R, Malik J (2015) Contextual action recognition with R-CNN. In: IEEE ICCV, pp 1080–1088
30. Ren S, He K, Girshick R, Sun J (2015) Faster R-CNN: towards real-time object detection with region proposal networks. In: Advances in neural information processing systems, pp 91–99
31. He K, Gkioxari G, Dollar P, Girshick R (2017) Mask R-CNN. In: ICCV, pp 2980–2988
32. Redmon J, Divvala S, Girshick R, Farhadi A (2016) You only look once: unified, real-time object detection. In: IEEE CVPR, pp 779–788
33. Molchanov VV, Vishnyakov BV, Vizilter YV, Vishnyakova OV, Knyaz VA (2017) Pedestrian detection in video surveillance using fully convolutional YOLO neural network. In: Automated visual inspection and machine vision II, vol 10334

34. Liu W, Anguelov D, Erhan D, Szegedy C, Reed S, Fu CY, Berg AC (2016) SSD: single shot multibox detector. In: European conference on computer vision, pp 21–37

35. Nie GH, Zhang P, Niu X, Dou Y, Xia F (2017) Ship detection using transfer learned single shot multi box detector. In: ITM web of conferences, vol 12, p 01006

36. He K, Zhang X, Ren S, Sun J (2016) Deep residual learning for image recognition. In: IEEE CVPR, pp 770–778

37. He K, Zhang X, Ren S, Sun J (2016) Identity mappings in deep residual networks. In: European conference on computer vision, pp 630–645

38. Goodfellow I, Pouget-Abadie J, Mirza M, Xu B, Warde-Farley D, Ozair S, Courville A, Bengio Y (2014) Generative adversarial networks. In: International conference on neural information processing systems (NIPS), pp 2672–2680

39. Shrivastava A, et al (2017) Learning from simulated and unsupervised images through adversarial training. In: CVPR'17

40. Mnih V et al (2015) Human-level control through deep reinforcement learning. Nature 518:529–533

41. Littman M (2015) Reinforcement learning improves behavior from evaluative feedback. Nature 521:445–451

42. Hasselt HV (2011) Double Q-learning. Adv Neural Inf Process Syst. 23:2613–2622

43. Cho K (2013) Simple sparsification improves sparse denoising autoencoders in denoising highly corrupted images. In: International conference on machine learning, pp 432–440

44. Zeng K, Yu J, Wang R, Li C, Tao D (2017) Coupled deep autoencoder for single image super-resolution. IEEE Trans Cybern 47(1):27–37

45. Xing C, Ma L, Yang X (2016) Stacked denoise autoencoder based feature extraction and classification for hyperspectral images. J Sens

46. Zamir A, et al (2018) Taskonomy: disentangling task transfer learning. In: CVPR'18

47. Hoo-Chang S, Roth HR, Gao M, Lu L, Xu Z, Nogues I, Summers RM (2016) Deep convolutional neural networks for computer-aided detection: CNN architectures, dataset characteristics and transfer learning. IEEE Trans Med Imag 35(5):1285

48. Li S (2009) Markov random field modeling in image analysis. Springer, Berlin

49. Koller D, Friedman N (2009) Probabilistic graphical models. MIT Press, Cambridge, MA

50. Detwarasiti A, Shachter RD (2005) Influence diagrams for team decision analysis. Decis Anal 2(4):207–228

51. Wu B, Iandola F, Jin PH, Keutzer K (2017) SqueezeNet: unified, small, low power fully convolutional neural networks for real-time object detection for autonomous driving. In: IEEE conference on computer vision and pattern recognition workshops, pp 129–137

52. Guan Y, Li C, Roli F (2015) On reducing the effect of covariate factors in gait recognition: a classifier ensemble method. IEEE Trans Pattern Anal Mach Intell 37(07):1521–1529

53. Veit A, Wilber MJ, Belongie S (2016) Residual networks behave like ensembles of relatively shallow networks. In: Advances in neural information processing systems, pp 550–558

54. De Boer PT, Kroese DP, Mannor S, Rubinstein RY (2005) A tutorial on the cross-entropy method. Ann Operat Res 134(1):19–67

55. Dunne RA, Campbell NA (1997) On the pairing of the softmax activation and cross-entropy penalty functions and the derivation of the softmax activation function. In: Australian Conference on the Neural Networks, Melbourne, vol 181, p 185

56. Cover T, Thomas J (1991) Elements of information theory. John Wiley & Sons Inc., Hoboken

57. Baeza-Yates R, Ribeiro-Neto B (2011) Modern information retrieval: the concepts and technology behind search, 2nd edn. Addison-Wesley, Boston, UK

58. Manning C, Raghavan P, Schutze H (2008) Introduction to Information Retrieval. Cambridge University Press, Cambridge

59. McCulloch WS, Pitts W (1943) A logical calculus of the ideas immanent in nervous activity. Bull Math Biophys 5(4):115–133

60. Itskov M (2011) Tensor algebra and tensor analysis for engineers, 4th edn. Springer, Berlin

61. Abadi M, Barham P, Chen J, Chen Z, Davis A, Dean J, Kudlur M (2016) TensorFlow: a system for large-scale machine learning. In: USENIX symposium on operating systems design and implementation (OSDI), USA, vol 16, pp 265–283
62. Muscat J (2014) Functional analysis. Springer, Berlin
63. Jacobson N (2009) Abstract algebra, 2nd Edn. Dover Publications, Mineola
64. LeCun Y, Ranzato M (2013) Deep learning tutorial. In: International conference on machine learning (ICML'13)
65. Schmidhuber J (2015) Deep learning in neural networks: an overview. Neural Netw 61:85–117
66. Kim Y (2014) Convolutional neural networks for sentence classification. In: Conference on empirical methods in natural language processing, pp 1746–1751
67. Liu Z, Yan WQ, Yang ML (2018) Image denoising based on a CNN model. In: International conference on control, automation and robotics (ICCAR), pp 389–393
68. Liu Z (2018) Comparative evaluations of image encryption algorithms. Masters thesis, Auckland University of Technology, Auckland
69. Ren Y (2017) Banknote recognition in real time using ANN. Masters thesis, Auckland University of Technology, Auckland, New Zealand
70. Wang H (2018) Real-time face detection and recognition based on deep learning. Masters thesis, Auckland University of Technology, Auckland
71. Zhang Q (2018) Currency recognition using deep learning. Masters thesis, Auckland University of Technology, Auckland, New Zealand
72. Xin C (2018) Detection and recognition for multiple flames using deep learning. Masters thesis, Auckland University of Technology, Auckland, New Zealand
73. Al-Sarayreh M (2020) Hyperspectral imaging and deep learning for food safety assessment. PhD thesis, Auckland University of Technology, Auckland, New Zealand
74. Al-Sarayreh M, Reis M, Yan W, Klette R (2019) A sequential CNN approach for foreign object detection in hyperspectral images. In: CAIP'19, pp 271–283
75. Cui W (2014) A scheme of human face recognition in complex environments. Masters thesis, Auckland University of Technology, Auckland, New Zealand
76. Wang X, Yan W (2020) Multi-perspective gait recognition based on ensemble learning. Springer Neural Comput Appl 32:7275–7287
77. Song C, He L, Yan W, Nand P (2019) An improved selective facial extraction model for age estimation. In: IVCNZ'19
78. Lu J (2016) Empirical approaches for human behavior analytics. Masters thesis, Auckland University of Technology, Auckland, New Zealand
79. An N (2020) Anomalies detection and tracking using siamese neural networks. Master thesis, Auckland University of Technology, Auckland, New Zealand
80. Wang X, Yan W (2019) Gait recognition using multichannel convolution neural networks. Springer neural computing and applications
81. Wang X, Yan W (2020) Human gait recognition based on frame-by-frame gait energy images and convolutional long short term memory. Int J Neural Syst 30(1):1950027:1–1950027:12
82. Wang X, Yan W (2019) Human gait recognition based on SAHMM. IEEE/ACM Trans Biol Bioinf
83. Liu C, Yan W (2020) Gait recognition using deep learning. In: Handbook of research on multimedia cyber security (IGI Global), pp 214–226
84. Li R (2017) Computer input of morse codes using finger gesture recognition. Masters thesis, Auckland University of Technology, Auckland, New Zealand
85. Zhang Y (2016) A virtual keyboard implementation based on finger recognition. Masters thesis, Auckland University of Technology, Auckland, New Zealand
86. Zheng K, Yan WQ, Nand P (2018) Video dynamics detection using deep neural networks. IEEE Trans Emerg Topics Comput Intell 2(3):224–234
87. Shen Y, Yan W (2018) Blindspot monitoring using deep learning. In: IEEE IVCNZ'18

88. Shen D, Chen X, Nguyen M, Yan WQ (2018) Flame detection using deep learning. In: International conference on control, automation and robotics (ICCAR), pp 416–420

89. Zhang Q, Yan W, Kankanhalli M (2019) Overview of currency recognition using deep learning. J Bank Financ Technol 3(1):59–69

90. Ma X (2020) Banknote serial number recognition using deep learning. Masters thesis, Auckland University of Technology, Auckland, New Zealand

91. Ji H, Liu Z, Yan W, Klette R (2019) Early diagnosis of Alzheimer's disease using deep learning. In: ICCCV'19, pp 87–91

92. Ji H, Liu Z, Yan W, Klette R (2019) Early diagnosis of Alzheimer's disease based on selective kernel network with spatial attention. In: ACPR'19, pp 503–515

93. Sun S (2020) Empirical analysis for earlier diagnosis of alzheimer's disease using deep learning. Masters thesis, Auckland University of Technology, Auckland, New Zealand

94. Vaswani A, et al (2017) Attention is all you need. In: The conference on neural information processing systems (NIPS), USA

95. Fu Y (2020) Fruit freshness grading using deep learning. Masters thesis, Auckland University, Auckland, New Zealand

96. Al-Sarayreh M, Reis M, Yan W, Klette R (2018) Detection of red-meat adulteration by deep spectral-spatial features in hyperspectral images. J Imag 4(5):63

97. Al-Sarayreh M, Reis M, Yan W, Klette R (2020) Potential of deep learning and snapshot hyperspectral imaging for classification of species in meat. Food Control 117:107332

98. Huang G, Liu Z, Weinberger KQ, van der Maaten L (2017) Densely connected convolutional networks. In: IEEE CVPR, vol 1, no 2, p 3

99. Rumelhart DE, Hinton GE, Williams RJ (1986) Learning representations by backpropagating errors. Nature 323(6088):533–536

100. Webb S (2018) Deep learning for biology. Nature 554:555–557

101. Zhu B, et al (2018) Image reconstruction by domain-transform manifold learning. Nature 555:487–492

102. George D et al (2017) A generative vision model that trains with high data efficiency and breaks text-based CAPTCHAs. Science 358(6368):eaag2612

103. Jordan MI, Mitchell TM (2015) Machine learning: trends, perspectives, and prospects. Science 349(6245):255–260

Deep Learning Platforms

2

2.1 Introduction

There are many deep learning platforms available such as Caffe (Fig. 2.1), Tensorflow, MXNet, Torch, and Theano. Caffe (Convolutional Architecture for Fast Feature Embedding) is a deep learning framework, which originally was developed at the University of California, Berkeley. Caffe supports visual object detection and classification as well as image segmentation using CNN, R-CNN, LSTM, and fully connected neural networks. Caffe supports GPU-based and CPU-based acceleration. Caffe2 includes new features such as Recurrent Neural Networks. At the end of March 2018, Caffe2 was merged into PyTorch.

PyTorch is an open-source machine learning library for applications such as computer vision and natural language processing. It was primarily developed by Facebook's AI Research Lab (FAIR). PyTorch defines a class called Tensor to store and operates on homogeneous multidimensional rectangular arrays of numbers.

MXNet was adopted by Amazon Web Services which supports lots of programming languages, such as C++, Python, R, Julia, etc. MXNet is a flexible and ultra-scalable deep learning framework that contains the state-of-the-art technology in deep learning, including convolutional neural networks (CNNs) and long short-term memory networks (LSTMs). MXNet al.so can be applied to both imperative and symbolic programming.

Theano is a Python-based optimizing compiler for mathematical expressions, especially matrix-valued one, which was developed by a Montreal Institute for Learning Algorithms (MILA). Theano is one of the most stable libraries, which allows automatic function gradient computations with its Python interface.

Usually Python includes the libraries such as NumPy (N-dimensional array package), SciPy (fundamental library for scientific computing), Matplotlib (comprehensive 2D plotting), Scikit-learn (machine learning library), etc.

© The Author(s), under exclusive license to Springer Nature Switzerland AG 2021
W. Q. Yan, *Computational Methods for Deep Learning*, Texts in Computer Science,
https://doi.org/10.1007/978-3-030-61081-4_2

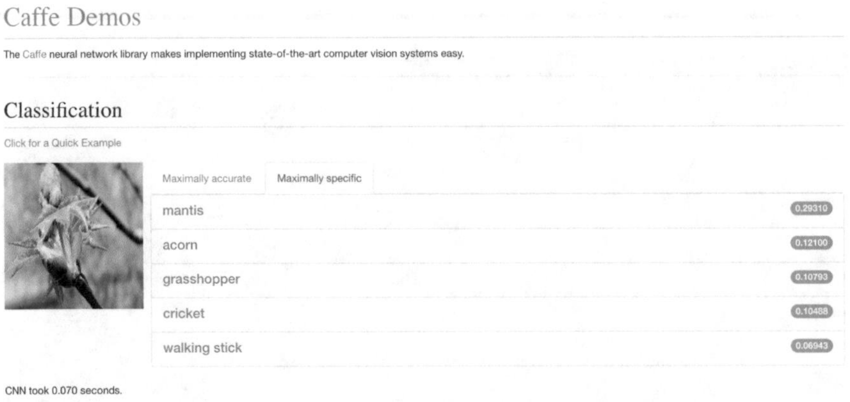

Caffe Demos

The Caffe neural network library makes implementing state-of-the-art computer vision systems easy.

Classification

Click for a Quick Example

Maximally accurate	Maximally specific	
mantis		0.29310
acorn		0.12100
grasshopper		0.10793
cricket		0.10488
walking stick		0.06943

CNN took 0.070 seconds.

Fig. 2.1 Visual object classification using Caffe

NumPy is the fundamental package for scientific computing with Python. Besides its obvious scientific uses, NumPy can also be used as an efficient multidimensional container of generic data. Arbitrary data types can be defined. This allows NumPy to seamlessly and speedily integrate with a wide variety of databases.

Matplotlib is a plotting library for Python programming and its mathematical extension, which is a platform for data visualization. According to its official website, the platform is a comprehensive library for creating static, animated, and interactive visualizations in Python. Matplotlib makes hard things possible.

Scikit-learn (scikits.learn or sklearn) is a free machine learning library for Python programming which features various classification, regression, and clustering algorithms including support vector machines (SVM), random forests, gradient boosting, k-means, etc. It also provides various tools for data fitting, preprocessing, model selection and evaluation based on NumPy, SciPy, and Matplotlib, etc.

2.2 MATLAB for Deep Learning

MATLAB is a multi-paradigm numerical computing environment and proprietary programming language developed by MathWorks. MATLAB allows matrix manipulations, plotting of functions and data, implementation of algorithms, creation of user interfaces, and interfacing with programs written in other languages.

The MathWorks logo is a surface plot of a variant of an eigenfunction of the L-shaped membrane. If $t \in (0, \infty)$ is time, $x \geq 0$ and $y \geq 0$ are the spatial coordinates with the units chosen so that the wave propagation speed is equal to one, then the amplitude of a wave satisfies the partial differential equation (2.1)

$$\frac{\partial^2 u}{\partial t^2} = \frac{\partial^2 u}{\partial x^2} + \frac{\partial^2 u}{\partial y^2}. \tag{2.1}$$

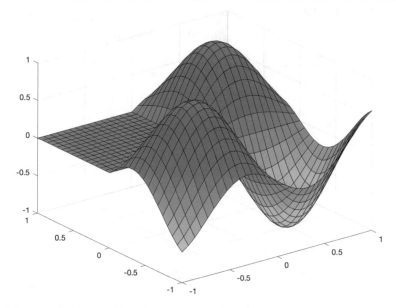

Fig. 2.2 The MATLAB membrane function with the parameter $k = 3$

Periodic time behaviour gives solutions of the form

$$u(t, x, y) = sin(y\sqrt{t})v(x, y), \tag{2.2}$$

where

$$\frac{\partial^2 v}{\partial x^2} + \frac{\partial^2 v}{\partial y^2} + \lambda v = 0, \tag{2.3}$$

where λ is the eigenvalue and the corresponding functions $v(x, y)$ are the eigenfunctions.

In MATLAB, the membrane function as shown in Fig. 2.2 is used to generate the MATLAB logo. $L = membrane(k)$, $k = 1, 2, \ldots, 12$ is the kth eigenfunction of the L-shaped membrane.

MATLAB utilizes toolboxes and the command line window to run programs. Especially, deep learning toolbox has been added into MATLAB in 2017. In 2019, MATLAB can build generative adversarial network (GAN), Siamese networks, variational autoencoders, and attention networks. MATLAB deep learning toolbox also can combine CNN and LSTM layers and networks that include 3D CNN layers [1].

MATLAB has its online version as shown in Fig. 2.3. The interfaces of MATLAB online version and offline version are almost the same. If the license is authorized and available, it is very convenient to access the software system and generate results. MATLAB provides demonstrations, documents, and source codes for further development [1].

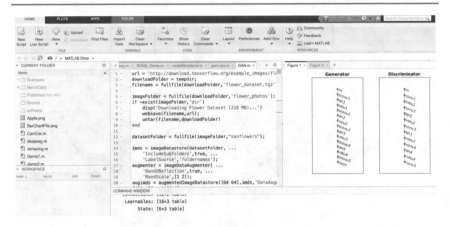

Fig. 2.3 The interface of MATLAB Online

MATLAB provides ANN (artificial neural network) toolbox in the early days which tells us how to deal with real applications, usually function fitting (nftool), pattern classification (nprtool), clustering (nctool), time series prediction and modelling (ntstool), etc.

Generally, we need to collect training data as a basic step, configure a network, initialize weights, and train the network; we need to minimize the differences, validate the classification. The confusion matrix is used to evaluate the results. ROC (receiver operating characteristic curve) and AUC (area under the curve) are calculated based on the classification. The ROC curve is plotted with TPR against FPR where TPR is on y-axis and FPR is on x-axis.

$$TPR = \frac{TP}{TP + FN}, \tag{2.4}$$

where TP is the true positive, FN is the false negative.

$$FPR = \frac{FP}{TN + FP}, \tag{2.5}$$

where FP is the false positive, TN is the true negative.

An excellent model has AUC near to 1.0 which means it has a good measure of separability. A poor model has AUC near to 0 which means it has the worst measure of separability.

Deep learning toolbox provides a framework for designing and implementing deep neural networks with algorithms, pretrained models, and APPs. MATLAB has the toolbox since 2017 including transfer learning, LSTM network for time series analysis, etc. The latest version includes AlexNet, GoogleNet, VGG-16/VGG-19, ResNet 101, Inception v2, generative adversarial network (GAN), reinforcement learning, etc. MATLAB can use multiple GPUs, parallel computing, cluster computing, cloud computing, etc., for accelerating deep learning processes.

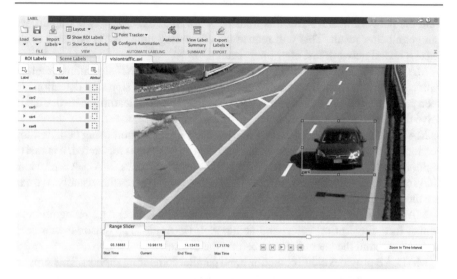

Fig. 2.4 MATLAB video labeler

MATLAB could be applied to time series analysis and forecast. Time series analysis comprises methods for analysing time series data in order to extract meaningful statistics and other characteristics of the data. Time series forecasting is the use of a model to predict future values based on previously observed data. Typical methods for time series analysis include spectral analysis, wavelet analysis, autocorrelation, and cross-correlation analysis. MATLAB provides Autoregressive (AR) and Autoregressive Integrated Moving Average models (ARIMA), and state-space models. In MATLAB deep learning toolbox, LSTM network as a kind of RNNs (Recurrent neural networks) has been applied to time series analysis and natural language processing (NLP) (NLP), which could help us to write or correct our writings.

MATLAB provides computer vision toolbox, especially for autonomous vehicles, visual object detection, semantic segmentation, digital image processing, etc. MATLAB also has a software Image Labeller for training data that could reduce our human labour. The Image Labeller and Video Labeller provide an easy way to mark the rectangular region of interest (ROI) labels, polyline ROI labels, pixel ROI labels, and scene labels in a video or image sequence. The video labeller automatically labels across image frames using an automation algorithm, e.g., the Kanade–Lucas–Tomasi (KLT) algorithm based on point tracking [2] as shown in Fig. 2.4. Following the steps (loading images, ROIs, labelling, data augmentation, exporting results, etc.), we can label all sampled images. ROIs (region of interest) will be marked and output for training and classification. This will be used to tell computer what the objects are in a scene [3].

MATLAB provides the function of transfer learning. That means, a network, e.g., AlexNet has been well trained, we can use the well-trained parameters first and apply it to a new network. We need to load the pretrained network, replace the final layers, train the network again. After this transfer, if we train the new network again, we can get a better result. This will reduce the computing time. MATLAB also could make a neural network fast after optimization.

MATLAB has embedded the Fast R-CNN and Faster R-CNN algorithms (regions with convolutional neural networks) already, an example for stop sign detection has been provided, the 11 lines source code as the simplest deep learning network could solve the specified task [4, 5].

MATLAB at present can run most of deep learning algorithms using both desktop version and online version as shown in Fig. 2.3. If an account is registered, it is easi to login. A MATLAB user can even develop their own toolboxes. MATLAB provides GUI for users to easy interactions. MATLAB can show our results visually. We can use the GUI interface to develop our applications.

MATLAB deep learning can be employed for biometrics, e.g., recognition of human face, fingerprint, voice, and aging, gait, DNA, etc. The reason is that deep learning can find the latent patterns behind the dataset.

MATLAB provides cloud computing and parallel computing support. This support could be used for face detection, visual object detection, vehicle detection, lane detection, and pedestrian detection, etc.

In the latest version, MATLAB supports artificial intelligence, event-based modelling, etc. In the latest version of deep learning, MATLAB users can now build generative adversarial networks (GANs), Siamese networks, variational autoencoders, and attention networks.

2.3 TensorFlow for Deep Learning

TensorFlow (https://www.tensorflow.org/) is a platform developed by Google and has been applied to deep learning. TensorFlow is run on a desktop (MS Windows, Mac OS, Linux, etc.) or online Colaboratory (Colab) (Colab) as shown in Fig. 2.5. Colab can provide GPU services online and runs entirely in the cloud. Through Colab, we can write and execute codes, save and share our experience, develop web, and access powerful computing resources all from a browser, without complicated configurations.

Tensor is a generalization of vectors and matrices to potentially higher dimensions. Generally speaking, in a tensor, the elements of a vector or matrix are still scalars, vectors, or matrices. TensorFlow is a framework to define and run computations involving tensors.

TensorFlow was especially developed for big data processing and visualization with TensorBoard [6] using graphs, numerical methods could be found from Tensor-Flow. TensorFlow presents tensors as n-dimensional arrays using base data types. These types reveal the relationships between different datasets.

TensorFlow not only has the normal data types but also includes special types, such as shape, variable, constant, placeholder, etc. The concept rank refers to a mathematical entity such as scalar, vector, matrix, etc.

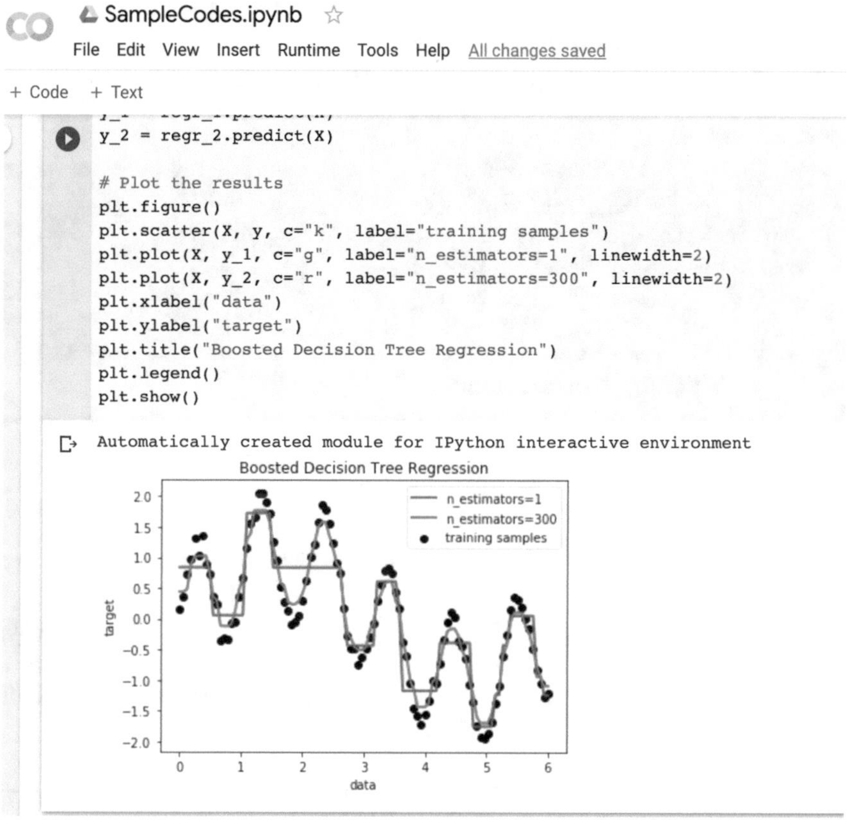

Fig. 2.5 Google colaboratory

The installation of TensorFlow is based on Mac OS/Unix, Microsoft Windows, Ubuntu, etc. After Python 3.0, pip3 is used to install Python-based applications.

C:> pip3 install - - upgrade tensorflow

TensorFlow needs a session to show the output, usually working along with the print command together to show the output of variables, for example, the famous "Hello, TensorFlow!" program is shown in Fig. 2.6.

A TensorFlow session encapsulates the state of runtime and operations. A session represents a connection between the client program, which accesses to hardware devices from the local machine and remote devices using the distributed TensorFlow runtime.

We have the exemplar source code to show the instance "add", "multiply", "dot product", "zero", etc. The "optimizer" could help us to quickly find a proper gradient using weights or variables from SGD (Stochastic Gradient Descent) algorithms. For

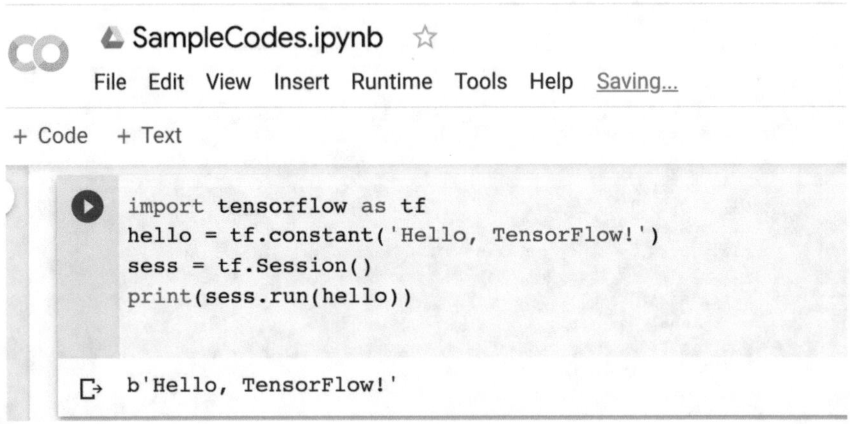

CO   📁 SampleCodes.ipynb ☆
      File   Edit   View   Insert   Runtime   Tools   Help   Saving...

+ Code   + Text

```
import tensorflow as tf
hello = tf.constant('Hello, TensorFlow!')
sess = tf.Session()
print(sess.run(hello))
```

↳   b'Hello, TensorFlow!'

Fig. 2.6 Hello, TensorFlow!

example, if $z = x^2 + xy$, $x, y, z \in \mathcal{R}$, then the gradients are

$$\begin{cases} \frac{\partial z}{\partial x} = 2x + y \\ \frac{\partial z}{\partial y} = x. \end{cases} \tag{2.6}$$

In order to minimize the function $z(\cdot)$, a standard gradient descent method would perform the following iterations or batches

$$\begin{cases} x' = x - \eta \cdot \frac{\partial z}{\partial x} \\ y' = y - \eta \cdot \frac{\partial z}{\partial y} \end{cases} \tag{2.7}$$

where η is a step size or the learning rate in machine learning. Provided $\eta = 0.1$, we randomly select $x = 5$, $y = 3$, using Eq. (2.7), then $x' = 3.7$ and $y' = 2.5$. We repeat this procedure, let $(x, y) \leftarrow (x', y')$ since $\eta < 1.0$, (x, y) will be converged and approximate to the local extreme point, namely,

$$\begin{cases} x_{n+1} = x_n - \eta \cdot \frac{\partial z}{\partial x_n} \\ y_{n+1} = y_n - \eta \cdot \frac{\partial z}{\partial y_n} \end{cases} \tag{2.8}$$

where $n = 1, 2, \ldots$, $\lim_{n \to \infty}(x_{n+1} - x_n) = 0$, $\lim_{n \to \infty}(y_{n+1} - y_n) = 0$, $\lim_{n \to \infty}(x_n, y_n) = (x_p, y_p)$, $\mathbf{P}(x_p, y_p)$ is the local extrem point.

TensorFlow graph is to show the computational network construction. The nodes (operations) and edges (tensors) indicate how individual operations are composed together. TensorFlow collections are stored by using metadata. TensorBoard is to render a computational graph through browsers like IE, Google Chrome, etc. TensorBoard is launched by using the following command line:

c:> tensorboard - - logdir="··· \tensorflow \ graph"

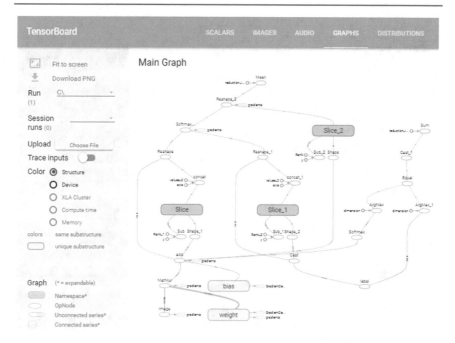

Fig. 2.7 A TensoprFlow graph of a network structure

Before that, we need to save the computational graph to a summary file using the function "tf.summary.FileWriter(·)". TensorBoard visualizes the structure of a graph in a browser under the support of a http server. The visualized result could be downloaded from the website: http://localhost:6006/ # graphs.

We show the graph of a neural network structure of a TensorFlow application as Fig. 2.7, the training accuracy of the TensorFlow application as shown in Fig. 2.8. We list two figures from TensorBoard as shown in Fig. 2.9 which reveal the dataset visualization using TensorFlow and TensorBoard.

The MNIST database is a large database of handwritten digits that is used for training various image processing systems. The MNIST database contains 60,000 training images and 10,000 test images. An extended dataset similar to MNIST called EMNIST has been published in 2017, which contains 240,000 training images and 40,000 test images of handwritten digits and characters.

For example, the steps of Convolutional Neural Network Estimator for MNIST (modified NIST) dataset are listed as

- **Step 1**. Load training and evaluation data.
- **Step 2**. Create the estimator/call the CNN model function.
- **Step 3**. CNN model function: convolutional layers, pooling layers, and fully connected layers.
- **Step 4**. Set up logging for predictions.
- **Step 5**. Train the model.

Fig. 2.8 Accuracy graph of a TensoprFlow training

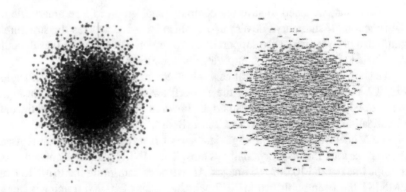

Fig. 2.9 TensoprFlow Graph Rendering (**a**) data samplse rendering, (**b**) data labels rendering

- **Step 6**. Evaluate the model and print results.

The steps of RNN routine using the MNIST dataset are listed as:

- **Step 1**. Set hyperparameters.
- **Step 2**. TensorFlow graph input.
- **Step 3**. Define weights.
- **Step 4**. Run RNN model function.
- **Step 5**. Hidden layer for output as the final results.

2.4 Data Augmentation

Image augmentation is a set of processing options of digital images, such as cropping, resizing, rotating, stretching and shearing, flipping, and reflection as well as the artefacts such as lense distortions, adding noises, and blurs.

In our projects, two distinct forms of data augmentation have been applied to human face detection so as to generate image translations and horizontal reflections, alter the intensities of the RGB channels in training images [7]. In the project currency recognition [8] and flame detection [9], image processing operations were taken into account such as scaling to the uniform size, clipping or expanding, cropping, random rotation, and colour adjustment. In the project of banknotes serial number recognition [10], the image augmentation approaches include image rotation, translation, colour jittering, and adding Gaussian noises.

The colour jittering enables us to alter the colours of an image by applying a random colour variation. For example, we can specify the range of hue, saturation, and gain (HSV) for random colours. We also can calculate principal components by using PCA algorithm of each colour matrix of an image and generate new variations by adding offset to the principal components. An example of colour jittering is shown in Fig. 2.10.

(a)

(b)

Fig. 2.10 A result of colour jittering using PCA algorithm, (**a**) original image, (**b**) image after colour jittering

In the project anomalies detection and object tracking [11], the data augmentation includes geometric transformation, Affine transformation, noise injection and random erasing, and so on. In the project vehicle-related scene understanding [3], offline augmentation and online augmentation are regarded as the two categories of data enrichment. The online augmentation includes rotations, translations, flipping, etc. The offline augmentation is typically employed for small datasets, which increases the size of the dataset by using a factor which equals to the number of conversions performed. In most of the cases, multiple transformation methods are amalgamated together to achieve much comprehensive expansion.

In the project fruit freshness grading [12], the image augmentation includes image scaling, rotating, cropping, and adding random noises based on observations. For adding random noises, the sequential order is random brightness adjustment, random contrast, and random erasion for digital images.

MATLAB provides image augmentation technologies by using the image processing Toolbox: Random image warping transformations, cropping transformations, colour transformations, synthetic noise, synthetic blur. An image data augmenter has been designed for a set of preprocessing options of image augmentation, such as resizing, rotation, and reflection.

2.5 Fundamental Mathematics

MATLAB was designed primarily for numerical analysis, especially all variables in MATLAB are arrays or vectors. TensorFlow derives from the operations on multidimensional data arrays, which are referred to tensors. For better understanding the implementations of deep learning algorithms in MATLAB, we introduce the fundamental knowledge of mathematics related to MATLAB.

For any two real numbers, we have the rules such as associative and commutatative relationships regarding two real numbers. In analysis, we have infinity (positive infinity: $+\infty$, negative infinity: $-\infty$), we also define the operations: $\infty \pm \infty$, $\frac{\infty}{\infty}$, etc.

In real analysis, we talk about the concept set. Based on the sets of real numbers, we construct function mapping from one set onto another. A function is continous over an interval $[a, b]$ that $\lim_{x \to x_0} f(x) = f(x_0) = f(x_0^+) = f(x_0^-)$, $x, x_0, x_0^+, x_0^- \in [a, b]$, $f(x) \in C[a, b]$. If $f(x)$ and $g(x) \in C[a, b]$, then $f(x) \pm g(x) \in C[a, b]$, $f(x) \times g(x) \in C[a, b]$, $f(x) \div g(x) \in C[a, b]$.

The differentiable means

$$f'(x_0) = \lim_{x \to x_0} \frac{f(x) - f(x_0)}{x - x_0} = f'(x_0^+) = f'(x_0^-) = \frac{df(x)}{dx}\big|_{x=x_0}. \qquad (2.9)$$

If $f'(x)$ and $g'(x) \in C[a, b]$, then $f'(x) \pm g'(x) \in C[a, b]$, $f'(x) \times g'(x) \in C[a, b]$, $f'(x) \div g'(x) \in C[a, b]$.

In chain rule, if $f(x) = g(y)$, $y = h(x)$, then $f(x) = g(h(y))$,

$$\frac{\partial f(x)}{\partial x} = \frac{\partial g(y)}{\partial y} \cdot \frac{\partial y}{\partial x} = \frac{\partial g(y)}{\partial y} \cdot \frac{\partial h(x)}{\partial x}. \tag{2.10}$$

Regarding Tylor expansion, given $f(x) \in C[a, b]$, we have

$$f(x) = f(x_0) + f'(x_0)(x - x_0) + \frac{1}{2!} f^{(2)}(x - x_0)^2 + \cdots + \frac{1}{k!} f^{(k)}(x - x_0)^k + \cdots . \tag{2.11}$$

That means all continuous functions defined over $[a, b]$ could be converted to polynomials. Typically, $\lim_{x \to 0} \frac{sinx}{x} = 1$, $sinx \approx x$.

We interpolate a curve using the given support points. The typical polynomials are quadratic curves, cubic polynomials, spline functions, Bezier functions, etc. Pertaining to Lagrange interpolating function, we have a polynomial with the degree n,

$$f(x) = \sum_{i=0}^{n} \prod_{i=0; i \neq j}^{n} \frac{(x - x_i)}{(x_j - x_i)} \cdot y_i, \tag{2.12}$$

where (x_i, y_i), $y_i = f(x_i)$, $i = 0, 1, \ldots, n$.

A vector space is a set **V** satisfying the following axioms:

- $\mathbf{x} + \mathbf{y} = \mathbf{y} + \mathbf{x}$ (addition is commutative)
- $(\mathbf{x} + \mathbf{y}) + \mathbf{z} = \mathbf{x} + (\mathbf{y} + \mathbf{z})$ (addition is associative)
- $\exists$ a unique vector zero 0, such that $0 + \mathbf{x} = \mathbf{x}$, $\forall \mathbf{x} \in \mathbf{V}$.
- $\forall \mathbf{x} \in \mathbf{V}$, $\exists$ a unique vector $-\mathbf{x}$ such that $\mathbf{x} + (-\mathbf{x}) = 0$.

To every pair α (real number) and $\mathbf{x}$ (vector), $\exists$ a vector $\alpha\mathbf{x}$, called scalar product of α and $\mathbf{x}$, such that

- $\alpha(\beta\mathbf{x}) = (\alpha\beta)\mathbf{x}$ (multiplication by scalars is associative)
- $1\mathbf{x} = \mathbf{x}$
- $\alpha(\mathbf{x} + \mathbf{y}) = \alpha\mathbf{x} + \alpha\mathbf{y}$ (distributive with respect to vector addition)
- $(\alpha + \beta)\mathbf{x} = \alpha\mathbf{x} + \beta\mathbf{x}$, $\alpha, \beta \in \mathbf{R}$ and $\mathbf{x} \in \mathbf{V}$ (distributive with respect to scalar addition)

A vector space has the properties:

- A basis in a vector space **V** is a set $\mathbf{G} = \{\mathbf{g}_1, \mathbf{g}_2, \ldots, \mathbf{g}_n\} \subset \mathbf{V}$ of linearly independent vectors such that every vector in **V** is a linear combination of elements of **G**.
- A vector space **V** is finite-dimensional $\|\mathbf{V}\| < \infty$ if it has a finite basis $\mathbf{G} = \{\mathbf{g}_1, \mathbf{g}_2, \ldots, \mathbf{g}_n\}$, $n < \infty$.
- The dimension of a finite-dimensional vector space **V** is the number of elements in a basis $\{\mathbf{g}_1, \mathbf{g}_2, \ldots, \mathbf{g}_n\}$ of **V**, namely, $\|\mathbf{V}\| = n$.

- Let $\mathbf{G} = \{\mathbf{g}_1, \mathbf{g}_2, \dots, \mathbf{g}_n\}$ be a basis of an n-dimensional vector space $\mathbf{V}$. Then, $\mathbf{x} = \sum_{i=1}^{n} x^i \mathbf{g}_i$ is Einstein's summation convention, $\mathbf{x} \in \mathbf{V}$.

The scalar (inner) product is a real-valued function $\mathbf{x} \cdot \mathbf{y}$ of two vectors $\mathbf{x}$ and $\mathbf{y}$ in a vector space $\mathbf{V}$.

- $\mathbf{x} \cdot \mathbf{y} = \mathbf{y} \cdot \mathbf{x}$ (commutative rule)
- $\mathbf{x} \cdot (\mathbf{y} + \mathbf{z}) = \mathbf{x} \cdot \mathbf{y} + \mathbf{x} \cdot \mathbf{z}$ (distributive rule)
- $\alpha(\mathbf{x} \cdot \mathbf{y}) = (\alpha \mathbf{x}) \cdot \mathbf{y} = \mathbf{x} \cdot (\alpha \mathbf{y})$, $\forall \alpha \in \mathbf{R}$, $\forall \mathbf{x}, \mathbf{y}, \mathbf{z} \in \mathbf{V}$ (associative rule for the multiplication by a scalar)
- $\mathbf{x} \cdot \mathbf{x} \geq 0$, $\forall \mathbf{x} \in \mathbf{V}$, $\mathbf{x} \cdot \mathbf{x} = 0$ if and only if $\mathbf{x} = 0$.
- Euclidean length (also called norm) of a vector $\mathbf{x}$, $\|\mathbf{x}\|_1 = \sqrt{\mathbf{x} \cdot \mathbf{x}}$
- Two non-zero vectors $\mathbf{x}$ and $\mathbf{y}$ are called orthogonal $\mathbf{x} \perp \mathbf{y}$, if $\mathbf{x} \cdot \mathbf{y} = 0$.
- A basis $\mathbf{E} = \{\mathbf{e}_1, \mathbf{e}_2, \dots, \mathbf{e}_n\}$ of an n-dimensional Euclidean space $\mathbf{E}^n$ is orthonormal if $\mathbf{e}_i \cdot \mathbf{e}_j = \delta_{ij}$, $i, j = 1, 2, \dots n$.
$$\delta_{ij} = \delta^{ij} = \delta_j^i = \delta_i^j = \begin{cases} 1 \ \mathbf{x} = \mathbf{y} \\ 0 \ \mathbf{x} \neq \mathbf{y} \end{cases} \text{(Kronecker delta)}$$
- $\mathbf{e}_1 = \frac{\mathbf{x}_1}{\|\mathbf{x}_1\|}, \dots, \mathbf{e}_n = \frac{\mathbf{e}_n'}{\|\mathbf{e}_n'\|}$ is the Gram-Schmidt process, where $\mathbf{e}_n' = \mathbf{x}_n - (\mathbf{x}_n, \mathbf{e}_{n-1})\mathbf{e}_{n-1} \cdots - (\mathbf{x}_n, \mathbf{e}_1)\mathbf{e}_1$.

Let $\mathbf{G} = \{\mathbf{g}_1, \mathbf{g}_2, \dots, \mathbf{g}_n\}$ be a basis in n-dimensional Euclidean space $\mathbf{E}^n$, a basis $\mathbf{G}' = \{\mathbf{g}_1', \mathbf{g}_2', \dots, \mathbf{g}'^n\}$ is dual to basis $\mathbf{G}$ if $\mathbf{g}_i \cdot \mathbf{g}_j' = \delta_{ij}$, $i, j = 1, 2, \dots, n$. $\mathbf{g}_i$ are linearly independent, if $\sum \alpha_i \mathbf{g}_i = 0$, then $\alpha_i = 0$.

The length of vector $\mathbf{x}$ can thus be written by $\|\mathbf{x}\| = \sqrt{x_i \mathbf{g}_i \cdot x_j' \mathbf{g}_i'} = \sqrt{x_i} \cdot x_j' \cdot \delta_{ij}$
$= \sqrt{x_i} \cdot x_j'$. e.g., $\mathbf{G} = \{\mathbf{e}_1, \mathbf{e}_2, \mathbf{e}_3\} = \{\mathbf{e}_2 \times \mathbf{e}_3, \mathbf{e}_3 \times \mathbf{e}_1, \mathbf{e}_1 \times \mathbf{e}_2\}$

if $<\mathbf{x}, \mathbf{y}> = \mathbf{x} + i\mathbf{y}$, $i = \sqrt{-1}$, then $<\alpha + i\beta><\mathbf{x}, \mathbf{y}> = <\alpha \mathbf{x} - \beta \mathbf{y})(\beta \mathbf{x} + \alpha \mathbf{y}>$, $\mathbf{x}, \mathbf{y} \in \mathbf{E}^n$, $\mathbf{z} = \mathbf{x} + i\mathbf{y} \in \mathbf{C}^n$, $\mathbf{A}(\mathbf{x} + i\mathbf{y}) = \mathbf{A}\mathbf{x} + i(\mathbf{A}\mathbf{y})$, $\mathbf{A} \in \mathbf{L}^n$.

For a tensor space, let $\mathbf{L}^n$ be a set of all linear mappings of one vector into another within $\mathbf{E}^n$, $\mathbf{y} = \mathbf{A}\mathbf{x}$, $\mathbf{x}, \mathbf{y} \in \mathbf{E}^n$, $\mathbf{A} \in \mathbf{L}^n$:

- Tensor linearity: $\mathbf{A}(\mathbf{x} + \mathbf{y}) = \mathbf{A}\mathbf{x} + \mathbf{A}\mathbf{y}$, $\forall \mathbf{x}, \mathbf{y} \in \mathbf{E}^n$, $\forall \mathbf{A} \in \mathbf{L}^n$
- $\mathbf{A}(\alpha \mathbf{x}) = \alpha(\mathbf{A}\mathbf{x})$, $\forall \mathbf{x} \in \mathbf{E}^n$, $\forall \alpha \in \mathbf{R}$, $\forall \mathbf{A} \in \mathbf{L}^n$
- Product of a tensor by a scalar number: $(\alpha \mathbf{A})\mathbf{x} = \alpha(\mathbf{A}\mathbf{x}) = \mathbf{A}(\alpha \mathbf{x})$, $\forall \mathbf{x} \in \mathbf{E}^n$
- Sum of tensors: $(\mathbf{A} + \mathbf{B})\mathbf{x} = \mathbf{A}\mathbf{x} + \mathbf{B}\mathbf{x}$
- Negative tensor: $-\mathbf{A} = (-1)\mathbf{A}$
- Zero tensor: $\mathbf{0}\mathbf{x} = 0$, $\forall \mathbf{x} \in \mathbf{E}^n$.
- Addition commutative: $\mathbf{A} + \mathbf{B} = \mathbf{B} + \mathbf{A}$
- Addition associative: $\mathbf{A} + (\mathbf{B} + \mathbf{C}) = (\mathbf{A} + \mathbf{B}) + \mathbf{C}$
- Element $\mathbf{0}$: $\mathbf{0} + \mathbf{A} = \mathbf{A}$, $\mathbf{A} + (-\mathbf{A}) = \mathbf{0}$
- Multiplication by scalars is associative: $\alpha(\beta \mathbf{A}) = (\alpha\beta)\mathbf{A}$
- Element $\mathbf{1}$: $\mathbf{1}\mathbf{A} = \mathbf{A}$
- Multiplication by scalars is distributive with respect to tensor addition: $\alpha(\mathbf{A} + \mathbf{B}) = \alpha\mathbf{A} + \alpha\mathbf{B}$

- Multiplication by scalars is distributive with respect to scalar addition: $(\alpha + \beta)\mathbf{A} = \alpha\mathbf{A} + \beta\mathbf{A}$, $\alpha, \beta \in \mathbf{R}$, $\mathbf{A}, \mathbf{B} \in \mathbf{L}^n$
- Vector product in $\mathbf{E}^3$, i.e., $\mathbf{z} = \mathbf{x} \times \mathbf{y}$, $\mathbf{x}, \mathbf{y}, \mathbf{z} \in \mathbf{E}^3$
- Rotation tensor: $\mathbf{R}(a)$, $a \in \mathbf{E}^3$ and $\mathbf{R} \in \mathbf{L}^3$

  For tensor functions:

- Function continuity:

$$\lim_{t \to t_0} \mathbf{x}(t) = \mathbf{x}(t_0), \tag{2.13}$$

$$\lim_{t \to t_0} \mathbf{A}(t) = \mathbf{A}(t_0). \tag{2.14}$$

- Differentiable:

$$\frac{d\mathbf{x}(t)}{dt} = \lim_{s \to 0} \frac{\mathbf{x}(t+s) - \mathbf{x}(t)}{s}, \tag{2.15}$$

$$\frac{d\mathbf{A}(t)}{dt} = \lim_{s \to 0} \frac{\mathbf{A}(t+s) - \mathbf{A}(t)}{s}. \tag{2.16}$$

- Product of a scalar function with a vector- or tensor-valued function:

$$\frac{d}{dt}[u(t)\mathbf{x}(t)] = \frac{du}{dt}\mathbf{x}(t) + \frac{d\mathbf{x}}{dt}u(t), \tag{2.17}$$

$$\frac{d}{dt}[u(t)\mathbf{A}(t)] = \frac{du}{dt}\mathbf{A}(t) + \frac{d\mathbf{A}}{dt}u(t). \tag{2.18}$$

- Scalar product of two vector- or tensor-valued functions:

$$\frac{d}{dt}[\mathbf{x}(t) \cdot \mathbf{y}(t)] = \frac{d\mathbf{x}}{dt} \cdot \mathbf{y}(t) + \mathbf{x}(t) \cdot \frac{d\mathbf{y}}{dt}, \tag{2.19}$$

$$\frac{d}{dt}[\mathbf{A}(t) : \mathbf{B}(t)] = \frac{d\mathbf{A}}{dt} : \mathbf{B}(t) + \mathbf{A}(t) : \frac{d\mathbf{B}}{dt}. \tag{2.20}$$

- Composition of two tensor-valued functions:

$$\frac{d}{dt}[\mathbf{A}(t)\mathbf{B}(t)] = \frac{d\mathbf{A}}{dt}\mathbf{B}(t) + \mathbf{A}(t)\frac{d\mathbf{B}}{dt}. \tag{2.21}$$

- $\mathbf{A}\mathbf{a} = \lambda\,\mathbf{a}, \mathbf{a} \neq 0$, $\mathbf{b}\mathbf{A} = \lambda\mathbf{b}, \mathbf{b} \neq 0$, $\mathbf{a}, \mathbf{b} \in \mathbf{C}^n$, $\lambda \in \mathbf{C}$ and $\mathbf{A} \in \mathbf{L}^n$. λ is an eigenvalue of tensor $\mathbf{A}$, $g(\mathbf{A}) = \sum_{k=0}^{m} a_k \mathbf{A}^k$, then

$$g(\lambda) = \sum_{k=0}^{m} a_k \lambda^k. \tag{2.22}$$

- Chain rule:

$$\frac{d}{dt}\mathbf{x}[u(t)] = \frac{d\mathbf{x}}{du}\frac{du}{dt}, \tag{2.23}$$

$$\frac{d}{dt}\mathbf{A}[u(t)] = \frac{d\mathbf{A}}{du}\frac{du}{dt}. \tag{2.24}$$

- Chain rule for functions of several arguments:

$$\frac{d}{dt}\mathbf{x}[u(t), v(t)] = \frac{d\mathbf{x}}{du}\frac{du}{dt} + \frac{d\mathbf{x}}{dv}\frac{dv}{dt}, \tag{2.25}$$

$$\frac{d}{dt}\mathbf{A}[u(t), v(t)] = \frac{d\mathbf{A}}{du}\frac{du}{dt} + \frac{d\mathbf{A}}{dv}\frac{dv}{dt}, \tag{2.26}$$

$$\frac{d}{dt}[\mathbf{A}(t)\mathbf{B}(t)] = \frac{d\mathbf{A}}{dt}\mathbf{B}(t) + \frac{d\mathbf{B}}{dt}\mathbf{A}(t). \tag{2.27}$$

- Directional derivatives:

$$\mathbf{r} = (\theta_1, \cdots, \theta_n), \theta_i \in \mathbf{R}. \tag{2.28}$$

For scalar field:

$$\frac{d}{ds}\Phi(\mathbf{r} + s\mathbf{a}) = \text{grad}\Phi \cdot \mathbf{a}, \forall \mathbf{a} \in \mathbf{E}^n, s \in \mathbf{R} \tag{2.29}$$

For vector field:

$$\frac{d}{ds}\mathbf{x}(\mathbf{r} + s\mathbf{a}) = \text{grad}\mathbf{x} \cdot \mathbf{a}, \forall \mathbf{a} \in \mathbf{E}^n, s \in \mathbf{R} \tag{2.30}$$

For tensor field:

$$\frac{d}{ds}\mathbf{A}(\mathbf{r} + s\mathbf{a}) = \text{grad}\mathbf{A} \cdot \mathbf{a}, \forall \mathbf{a} \in \mathbf{E}^n, s \in \mathbf{R} \tag{2.31}$$

2.6 Questions

Question 1. Should we use PyChem or only IDE?

Question 2. How to utilize the source codes and datasets from the GitHub website?

Question 3. What's the relationship between deep learning and machine learning? What are the supervised learning and unsupervised learning?

Question 4. How to choose an algorithm effectively for car detection, pedestrian detection, rubbish detection?

Question 5. What is Mask R-CNN [5]?

Question 6. What is the relationship between AI and deep learning?

Question 7. How to deal with the relationship between papers and source codes in a project development?

Question 8. Why mathematics is so important in deep learning and AI?

References

1. Vedaldi A, Lenc K (2015) MatConvNet: convolutional neural networks for matlab. In: ACM international conference on multimedia, pp 689–692
2. Klette R (2014) Concise computer vision: an introduction into theory and algorithms. Springer, London, UK
3. Liu X (2019) Vehicle-related scene understanding using deep learning. Masters thesis, Auckland University of Technology, New Zealand
4. Ren S, He K, Girshick R, Sun J (2015) Faster R-CNN: towards real-time object detection with region proposal networks. In: Advances in neural information processing systems, pp 91–99
5. He K, Gkioxari G, Dollar P, Girshick R (2017) Mask R-CNN. In: ICCV, pp 2980–988
6. Abadi M, Barham P, Chen J, Chen Z, Davis A, Dean J, Kudlur M (2016) TensorFlow: a system for large-scale machine learning. In: USENIX symposium on operating systems design and implementation (OSDI), USA, vol 16, pp 265–283
7. Wang H (2018) Real-time face detection and recognition based on deep learning. Masters thesis, Auckland University of Technology, Auckland
8. Zhang Q (2018) Currency recognition using deep learning. Masters thesis, Auckland University of Technology, Auckland, New Zealand
9. Xin C (2018) Detection and recognition for multiple flames using deep learning. Masters thesis, Auckland University of Technology, Auckland, New Zealand
10. Ma X (2020) Banknote serial number recognition using deep learning. Masters thesis, Auckland University of Technology, Auckland, New Zealand
11. An N (2020) Anomalies detection and tracking using siamese neural networks. Master thesis, Auckland University of Technology, Auckland, New Zealand
12. Fu Y (2020) Fruit freshness grading using deep learning. Masters thesis, Auckland University, Auckland, New Zealand

CNN and RNN

3

3.1 CNN and YOLO

Since 2015, the focus of all researchers has been moved to deep learning, i.e., deep neural networks, especially after AlexNet [1] received an award in a contest of visual object detection and recognition using ImageNet [2–4]. In 2015, the world top journal Nature also published a survey paper related to deep learning [5]. Before that, most of the people were interested in using SVM (support vector machine) for pattern classification [6, 7].

CNN (convolutional neural network or ConvNet) has been employed to digital image processing since 1995 [8]. The convolutional kernels usually are the masks with the size of $3 \times 3, 5 \times 5, 7 \times 7, 9 \times 9$, etc. The convolution operations generate receptive fields which compose the feature map of convolutional neural networks [4, 9–11]. The receptive field corresponds to a region of the origial image [12].

In mathematics, convolution is a mathematical operation on two functions that produces a third function expressing how the shape of one is modified or filtered by the other. Given $\mathbf{H} = (h_{i,j}^{(k)})_{m \times m}$ at level k, $a^{(k)}, b^{(k)}, c^{(k)}, d^{(k)} \in \mathcal{R}$, $g(\cdot)$ is a nonlinear function, a convolution operation is

$$h_{i,j}^{(k+1)} = g(a^{(k)} \cdot h_{i,j}^{(k)} + b^{(k)} \cdot h_{i+1,j}^{(k)} + c^{(k)} \cdot h_{i,j+1}^{(k)} + d^{(k)} \cdot h_{i+1,j+1}^{(k)}). \quad (3.1)$$

Average pooling includes calculating the average for each patch of the feature map. For an average pooling [13] with downsampling,

$$\bar{h}^{(k+1)} = \frac{1}{4}(a^{(k)} \cdot h_{i,j}^{(k)} + b^{(k)} \cdot h_{i+1,j}^{(k)} + c^{(k)} \cdot h_{i,j+1}^{(k)} + d^{(k)} \cdot h_{i+1,j+1}^{(k)}). \quad (3.2)$$

For a max pooling [14] with downsampling,

$$h_{\max}^{(k+1)} = \max(a^{(k)} \cdot h_{i,j}^{(k)}, b^{(k)} \cdot h_{i+1,j}^{(k)}, c^{(k)} \cdot h_{i,j+1}^{(k)}, d^{(k)} \cdot h_{i+1,j+1}^{(k)}), \quad (3.3)$$

© The Author(s), under exclusive license to Springer Nature Switzerland AG 2021
W. Q. Yan, *Computational Methods for Deep Learning*, Texts in Computer Science,
https://doi.org/10.1007/978-3-030-61081-4_3

where the max pooling is carried out by applying a max filter $\max(\cdot)$ to non-overlapping subregions of the initial representation. In deep learning, convolution operation '$\star$' is denoted as

$$s(t) = (x \star w)(t) = \sum_{a=-\infty}^{\infty} x(a)w(t-a), \qquad (3.4)$$

where the function $x(a)$ is input and $w(t)$ stands for kernel, and the output $s(t)$ represents feature map. For an image $I(i, j), i = 1, 2, \ldots, W, j = 1, 2, \ldots, H, W$ is the image width, H is the image height. The convolution operation is

$$S(i, j) = (I \star K)(i, j) = \sum_{m} \sum_{n} I(m, n)K(i-m, j-n), \qquad (3.5)$$

where $K(\cdot)$ is kernel function. Typically, Gaussian kernel in $n \in \mathscr{Z}^{+}$ dimensions is $G_n(\mathbf{X}, \sigma) = \frac{1}{(\sigma\sqrt{2\pi})^n} \exp(-\frac{\|\mathbf{X}\|^2}{2\sigma^2})$, where $\mathbf{X} = (x_1, x_2, \ldots, x_n)$, σ is variance. For example, if $n = 1$, $G_1(x, \sigma) = \frac{1}{\sigma\sqrt{2\pi}} \exp(-\frac{x^2}{2\sigma^2})$.

In convolution operation, padding is to fill up the region of an image boundary, which is to fill in the edge region with zero, this will make sure all convolution operations could be carried out at the edge region of images [13]. Meanwhile, the concept stride is the step length of convolution operations.

Convolution operation is to simulate our human visual system. Like most mammals such as cats or dogs, our human visual system (HVS) could be simulated by using the famous Gabor function. Gabor function has been applied to describe the co-occurence of texture in texture analysis [15].

$$G(x, y, \alpha, \beta_x, \beta_y, f, \phi, x_0, y_0, \tau) = \alpha \cdot \exp\left(-\beta_x {x'}^2 - \beta_y {y'}^2\right) \cdot \cos(fx' + \phi), \quad (3.6)$$

where $\alpha, \beta_x, \beta_y, f, \phi, x_0, y_0$, and τ are parameters,

$$x' = (x - x_0)\cos(\tau) + (y - y_0)\sin(\tau) \qquad (3.7)$$

$$y' = -(x - x_0)\sin(\tau) + (y - y_0)\cos(\tau). \qquad (3.8)$$

ConvNets (i.e., CNN) also include local connections, shared weights, pooling [13, 16], and multilayer neural network (MLP) [17, 18]. It is famous for its fine-tuning and pooling techniques [13].

3.1.1   R-CNN

The next is a region-based CNN (R-CNN). We need to understand the concept: Intersection over Union (IoU) for visual object detection in an image, which is calculated through

$$IoU = \frac{\mathscr{A}(A \bigcap B)}{\mathscr{A}(A \bigcup B)}, \qquad (3.9)$$

where anchor box refers to the box with multiresolution, multiscale, and multiaspectratio, etc.

R-CNN [19, 20] quickly finds the object box at where the features are extracted from ROI (region of interest), the classifier is still SVM, the regression is used for classifying region proposals iteratively. Warp refers to anisotropically scale each object proposal to the CNN input size. R-CNN is slow because it performs a ConvNet forward pass for each object proposal without sharing computations [21]. The training is a multistage pipeline; that means, we need work one step after another; therefore, it is expensive and time-consuming.

In MATLAB, R-CNN detector firstly generates region proposals. The proposal regions are cropped out of the image and resized. CNN classifies the cropped and resized regions. Finally, the region proposal bounding boxes are refined by using CNN features.

The training of Fast R-CNN [22, 23] is single stage through using a multitask loss. Regression has been employed for bounding box training [17, 24, 25]. The training can update all network layers. The pooling layer uses max pooling to convert the features inside any valid ROI into a small feature map with a fixed spatial extent of 7×7 region.

In MATLAB, Fast R-CNN deals with the entire image and pools CNN features corresponding to each region proposal. Generally speaking, Fast R-CNN is more efficient than R-CNN, which is the design purpose of this deep learning model.

Faster R-CNN [26, 27] merged Region Proposal Network (RPN) and Fast R-CNN into a single network by sharing their convolutional features. A RPN is a fully convolutional network that predicts object boundary and objectiveness scores at each position simultaneously.

The softmax function [28] is applied to object detection by using Faster R-CNN [26].

$$f(x) = \frac{e^{x_i}}{\sum_i e^{x_i}}, x \in (-\infty, \infty). \tag{3.10}$$

In MATLAB, Faster R-CNN adds a RPN to generate region proposals directly in the network. The RPN uses anchor boxes for visual object detection. Generating region proposals in the network is faster.

3.1.2 Mask R-CNN

Mask R-CNN is an intuitive extension of Faster R-CNN, constructing the mask branch properly is critical for generating good results [29].

Mask R-CNN extends Faster R-CNN by adding a branch for predicting segmentation masks on each Region of Interest (RoI), in parallel with the existing branch for classification and bounding box regression.

The mask branch is a small fully convolutional network (FCN) applied to each ROI, predicting a segmentation mask in a pixel-to-pixel manner. Mask R-CNN is simple to be implemented and trained given the Faster R-CNN framework, which facilitates a range of flexible architectures.

Mask R-CNN has been awarded as the best work in deep learning for its simple, flexible, and general framework for object instance segmentation. The team from Facebook AI Research have received the Best Paper Award (Marr Prize) at the 16th International Conference on Computer vision (ICCV) 2017, held in Venice, Italy.

3.1.3 YOLO

YOLO [30] (You Only Loook Once) is single-pass neural network, directly optimized the neural network; given an image, immediately only 7×7 segmentation is used for image segmentation; thus, YOLO is very fast.

YOLO network has 24 convolutional layers followed by 2 fully connected layers, which uses alternating 1×1 convolutional layers to reduce the feature space between layers. The convolutional layers are pretrained for classification by using the ImageNet dataset. YOLO adopts leaky rectified linear activation (ReLU) function $\phi(x) \in C^0(-\infty, \infty)$

$$\phi(x) = \begin{cases} x, & x > 0 \\ 0.1x, & \text{others} \end{cases} \tag{3.11}$$

YOLOv2 predicts the location and precision using logistic activation function (*a.k.a.* sigmoid function) $\sigma(\cdot)$. Namely,

$$f(x) = \sigma(x) = \frac{1}{1 + e^{-x}}, x \in C^\infty(0, 1). \tag{3.12}$$

The derivative of this monotonic function *w.r.t.* x,

$$f'(x) = f(x)(1 - f(x)), x \in C^\infty(0, 1). \tag{3.13}$$

YOLOv2 also uses 448×448 images for fine-tuning the classification network based on ImageNet. Batch normalization (BN) is based on all convolutional layers in YOLOv2. DarkNet-19 has 19 layers depth which can be trained based on more than a million images from the ImageNet database. The pretrained network can classify images into 1000 object categories. DarkNet-19 is often used as the foundation network for YOLO workflows.

YOLOv2 uses k-means clustering which leads to good IoU scores. Mathematically, k-means clustering partitions n observations into $k \leq n$ sets $S = \{S_1, S_2, \ldots, S_k\}$ so as to minimize the within-cluster sum of squares (WCSS). Namely,

$$S_K = \arg\min_S \sum_{i=1}^k \sum_{\mathbf{x} \in S_i} \|\mathbf{x} - \mu_i\|^2, \tag{3.14}$$

where μ_i is the mean of points in S_i.

In MATLAB, YOLOv2 object detector makes use of a single-stage object detection network and anchor boxes to detect classes of visual objects in an image. For each anchor box, YOLOv2 provides information such as IoU, anchor box offsets, and class probability.

YOLO9000 can detect over 9000 visual object categories using WordTree in real time [31]. WordTree has a hierarchical tree to link the classes and subclasses together. YOLO9000 provides a way to combine MS COCO and ImageNet together.

YOLOv3 used the Darknet, which has 53 layer network trained on ImageNet. YOLOv3 makes prediction at three scales, which are precisely given by downsampling the dimensions of the input image by 32, 16, and 8, respectively. YOLOv3 takes advantage of nine anchor boxes in total.

YOLOv4 is faster and more accurate than other real-time neural networks based on Microsoft COCO dataset. MSCOCO dataset includes three parts: Training set (120,000 images), validation set (5,000 images), test set (41,000 images). By using Darknet framework, YOLOv4 is able to cope with 62 frames with the resolution 608×608 per second and achieve 43.5% AP accuracy.

YOLOV4 is optimal and suitable for real-time object detection. Usually, the minimum speed is 30 FPS (frames per second) or more. For the resolution to detect multiple objects with various sizes and the exact location, a higher receptive field is required to keep more details of the visual objects.

3.1.4   SSD

SSD [32, 33] is single shot multibox detector(SSD). Single shot refers to the tasks of visual object localization and classification are done in a single forward pass of the network. MultiBox is the name of a technique for bounding box regression. The network is a visual object detector that also classifies those detected objects.

The architecture of SSD is based on the venerable VGG-16 architecture but discards the fully connected layers. A set of auxiliary convolutional layers were added and employed to extract features at multiple scales, which progressively decrease the size of the input to each subsequent layer. SSD makes use of the default aspect ratio and could be applied to visual object tracking in real time [34, 35].

There are a few implementations of SSD available online, including the original Caffe code. TensorFlow-based SSD code could be downloaded from GitHub.

3.1.5   DenseNets and ResNets

DenseNets alleviate the gradient varnishing problem, strengthen feature propagation, encourage feature reuse, and substantially reduce the number of parameters [36].

For each layer, the feature maps of all preceding layers are used as inputs, its own feature maps are used as inputs into all subsequent layers.

DenseNets introduce direct connections between any two layers with the same feature map size, which scale naturally to hundreds of layers, while exhibit no optimization difficulties. Thus, DenseNets require substantially fewer parameters and less computation to achieve state-of-the-art performances. DenseNets allow feature reuse throughout the networks consequently learn more compact and more accurate models.

In deep learning, there exists the degradation problem; namely, with the network depth increasing, accuracy gets saturated. However, ResNets easily obtain accuracy gains from greatly increased depth due to the well-designed structure of this network.

$$\mathbf{y} = \mathscr{F}(\mathbf{x}, \{W_i\}) + \mathbf{x}, \tag{3.15}$$

where $\mathbf{x}$ and $\mathbf{y}$ are the input and output vectors of the layers, $\mathscr{F}(\cdot)$ is the residual mapping, e.g., $\mathscr{F} = W_{2\sigma}(W_{1x})$, σ is the ReLU function.

3.2   RNN and Time Series Analysis

Recurrent neural network (RNN) is one of the deep neural networks which is now applied to many aspects widely since its unique structure is quite helpful and beneficial when dealing with sequence data.

RNN structure has the recurrent hidden layer connected to that of the next step. Compared with other multilayer neural networks, RNN can impact over time. To explain this in more details, there is a unidirectional flow of information from the input unit to the hidden unit, while there is another unidirectional flow of information from the hidden unit to the output one. In addition, the input of the hidden layer also contains the state of the previously hidden layer; the nodes of the hidden layer can be self-connected or interconnected.

An LSTM [37] network is a type of recurrent neural networks (RNN) that learns long-term dependencies between time steps of sequence data. The core components of an LSTM network are a sequence input layer and an LSTM layer. The input layer imports sequence or time series data into the network. LSTM networks can remember the state of the network.

In MATLAB, LSTM networks support input data with varying sequence lengths. When passing data through the network, the network pads, truncates, or splits sequences so that all the sequences in each mini-batch have the specified length.

If x is the input layer, o is the output layer, t is the number of times, s is the hidden layer, V, W, and U are all weights, the state of the hidden layer at time t can be calculated as

$$S_t = f(U \cdot x_t + W \cdot S_{t-1}), \tag{3.16}$$

where $f(\cdot)$ is the activation function.

If there is a sequence of inputs $x_1, x_2, \ldots, x_T \in \mathscr{R}^n$, the sequence of hidden states, calculated by the network, is $h_1, h_2, \ldots, h_T \in \mathscr{R}^m$, the sequence of prediction is $y_1, y_2, \ldots, y_T \in \mathscr{R}^k$, the following equations are employed for iterations:

$$t_i = W_h^x x_i + W_h^h x_{i-1} + b_h, \tag{3.17}$$

$$h_i = e(t_i), \tag{3.18}$$

$$s_i = W_y h h_i + b_y, \tag{3.19}$$

$$y_i = g(s_i), \tag{3.20}$$

where W_h^x, W_h^h, and W_y^h are the weight matrices; the sequence of t_i represents the inputs to the hidden units, the sequence of s_i stands for the inputs to the output units; b_h and b_y are bias vectors; $e(\cdot)$ and $g(\cdot)$ are the predefined vector-valued functions.

3.3 HMM

We usually use FSM and HMM (hidden Markov model) [38–41] to detect events. HMM is generally employed to predict what will happen there. A typical example is to predict whether a person is healthy or fever in a day using probability. It will be helpful for preparing medicine in a hospital for a season based on weather changes. The stochastic automation is:

(1) Initial probabilities: $\pi_i \equiv p(q_1 = S_i)$, $\sum_{i=1}^{N} \pi_i = 1$; $\Pi = (\pi_1, \pi_2, \ldots, \pi_N)$.

(2) Transition matrix: $\mathbf{A} = (a_{ij})_{N \times N}$, $a_{ij} \equiv p(q_{t+1} = S_j | q_t = S_i) \in [0, 1]$ and $\sum_{j=1}^{N} a_{ij} = 1$

HMM model $\lambda = (\mathbf{A}, \mathbf{B}, \Pi)$ is related to

1. State: $\mathbf{S} = \{S_1, S_2, \ldots, S_N\}$
2. Observation: $\mathbf{V} = \{v_1, v_2 \cdots, v_M\}$
3. Transition matrix: $\mathbf{A} = (a_{ij})_{N \times N}$, $a_{ij} \equiv p(q_{t+1} = S_j | q_t = S_i)$
4. Emission probabilities: $\mathbf{B} = (b_j(m))_M$, $b_j(m) \equiv p(O_t = v_m | q_t = S_j)$
5. Initial probabilities: $\Pi = (\pi_i)_N$, $\pi_i \equiv p(q_1 = S_i)$
6. Output: $\mathbf{O} = \{O_1 O_2 \cdots O_T\}$
7. Latent variables: $\mathbf{Q} = \{Q_1 Q_2 \cdots Q_T\}$

HMM has two very important algorithms: Viterbi algorithm and Baum–Welch (BM) algorithm. The Viterbi algorithm could help us quickly find the best path which has been applied to information theory for coding.

Given $\mathbf{Q} = \{q_1, \ldots, q_T\}$ and $\mathbf{O} = \{o_1, \ldots, o_T\}$,

$$\delta_t(i) = \max \ p(q_1 q_2 \cdots q_{t-1}, q_t = S_i, O_1 \cdots O_t | \lambda). \tag{3.21}$$

Computationally,

1. Initialization: $\delta_1(i) = \pi_i b_i(O_1)$, $\psi_1(i) = 0$
2. Recursion: $\delta_t(j) = \max_i (\delta_{t-1}(i) \cdot a_{ij}) b_j(O_t)$, $\psi_t(j) = \arg \max_i (\delta_{t-1}(i) \cdot a_{ij})$
3. Termination: $p^* = \max_i \delta_T(i)$, $q_T^* = \arg \max_i \delta_T(i)$
4. Path: $q_t^* = \psi_{t+1}(q_{t+1}^*)$, $t = T - 1, T - 2, \ldots, 1$

BM algorithm is applied to predict the highest probability of parameters by using EM algorithm. Given $\lambda = (\mathbf{A}, \mathbf{B}, \Pi)$, Baum–Welch (BW) algorithm is used to seek $\lambda^* = \arg \max_{\lambda} p(\chi | \lambda)$,

E-step,

$$\gamma_t(i) = \sum_{j=1}^{N} \xi_t(i, j); \xi_t(i, j) \equiv p(q_t = S_i, q_{t+1} = S_j | \mathbf{O}, \lambda); \qquad (3.22)$$

M-step,

$$p(\chi | \lambda) = \Pi_{k=1}^{K} p(O^k | \lambda) \qquad (3.23)$$

$$\hat{a}_{ij} = \frac{\sum_{t=1}^{T-1} \xi_t(i, j)}{\sum_{t=1}^{T-1} \gamma_t(i, j)}; \hat{b}_j(m) = \frac{\sum_{t=1}^{T} \gamma_t(j)\mathbf{1}(O_t = v_m)}{\sum_{t=1}^{T} \gamma_t(j)}. \qquad (3.24)$$

HMMs use transition probability in each step for predicting events what will be happened. FSM has no probability prediction, which is used to capture events during state transitions.

HMM is not a neural network, it has not neurons and activation functions. RNNs [42–44] are a family of artificial neural networks for processing sequential data, which is a dynamical system driven by external $x^{(t)}$,

$$h^{(t)} = f(h^{(t-1)}, x^{(t)}; \theta) = g^{(t)}(x^{(t)}, x^{(t-1)}, \ldots, x^{(1)}), \qquad (3.25)$$

where $t = 1, 2 \ldots, \tau$, h is the state. This tells us it is possible to use the same transition function with the same parameters at every time step unfolding.

3.3.1   RNN: Recurrent Neural Networks

RNN [43–46] refers to recurrent neural network; most of the time, we unfold the neural network, the procedure is simulated usually by using the fixed-point theorem. We calculate the difference using loss functions. A loss function is a part of a cost function which is a type of an objective function. Hereinafter, the concepts, loss function, cost function, and objective function, have minor differences.

Loss function $L(y_i, \hat{y}_i)$ refers to a single sample in a dataset, $\hat{y}_i$ is the output of a nerual network model, y_i is the real value or ground truth. Cost function $J(\cdot)$ means the entre training dataset with all samples $J = \sum L(y_i, \hat{y}_i), i = 1, 2, \ldots, n$, for example, mini-batch in gradient descent uses all the samples of the training set. Objective function means a function $f(\cdot)$ will be optimized by using optimization algorithm which is subject to constraints.

In mathematics, the basic concept of loss function [47] is a distance. The popular one is Euclidean distance, but entropy $e = -\sum_{i=1}^{n} h_i \log h_i = -\mathbf{E}(\log h_i), h_i \in (0, 1]$ has been applied to calculate the distance using mathematical expectation of a logarithm function.

The softmax function $f(x) = \frac{e^{x_i}}{\sum_i e^{x_i}}$ has been used to the calculation [28, 48, 49].

The loss functions typically have $0 \sim 1$ loss function, square loss function, absolute loss function, average loss function, hinge loss function, etc. $0 \sim 1$ loss function is

$$L(Y, f(X)) = \begin{cases} 1 & Y \neq f(X) \\ 0 & Y = f(X) \end{cases} \tag{3.26}$$

Squared error cost function or quadratic cost function is shown as

$$J = \|\mathbf{Y}, f(\mathbf{X})\|^2 = \sum_{i=1}^{n} (y_i - f(x_i))^2, \tag{3.27}$$

where (x_i, y_i), $i = 1, 2 \ldots, n$ is a group of given points, $\mathbf{X} = (x_1, x_2, \ldots, x_n)^\tau$, $Y = (y_1, y_2, \ldots, y_n)^\tau$, y_i is different from $f(x_i)$.

The squared cost function has an important position in linear algebra. For example, for a straight line $y = ax + b$, where parameters a and b are unknown, if you have n 2D sampling points $\mathbf{p} = (x_i, y_i)^\tau$, $i = 1, 2, \ldots, n$, we can treat $\theta = (a, b)^\tau$ as parameters, which can be estimated by using linear regression. Hence,

$$J(a, b) = \sum_{i=1}^{n} (ax_i + b - y_i)^2. \tag{3.28}$$

We rewrite the Eq. (3.28) as a quadratic polynomial, a quadratic polynomial is a polynomial of degree 2, a bivariate quadratic polynomial has the form

$$J(a, b) = A \cdot a^2 + B \cdot b^2 + C \cdot ab + D \cdot a + E \cdot b + F, \tag{3.29}$$

where A, B, C, D, E, and F are the constants ($A \neq 0$) which are related to (x_i, y_i), $n = 1, 2, \ldots, n$. Bivariate polynomials are fundamental to the study of conic sections, which are characterized by equating the expression $L(a, b)$ to zero. Thus, we modify the Eq. (3.29) and obtain

$$J(a, b) = a^2 + \frac{B}{A} \cdot b^2 + \frac{C}{A} \cdot ab + \frac{D}{A} \cdot a + \frac{E}{A} \cdot b + \frac{F}{A}, A \neq 0. \tag{3.30}$$

Moreover, we simplify the Eq. (3.30) in the form of matrix

$$J(a, b) = (a, b, 1)\mathbf{M}(a, b, 1)^\tau, \tag{3.31}$$

where $\mathbf{M} = (m_{i,j})_{3 \times 3}$, $m_{i,j}$ was derived from the constants A, B, C, D, E, and F. We see that matrices could be applied to express a quadratic polynomial, linear algebra could be applied to the square loss function.

Absolute loss function is experssed as

$$L(Y, f(X)) = |Y - f(X)|. \tag{3.32}$$

Logarithm loss function is

$$L(Y, p(Y|X)) = -\log p(Y|X), \tag{3.33}$$

where $p(Y|X)$ is the conditional probability. The average cost function is

$$\bar{J} = \frac{1}{m} \sum_{i=1}^{m} L(x_i, y_i), \tag{3.34}$$

where the set $T = \{(x_i, y_i)\}(i = 1, 2, \ldots, m)$ is the training dataset, $\mathbf{X} = (x_1, x_2, \ldots, x_m)$, $\mathbf{Y} = (y_1, y_2, \ldots, y_m)$.

In machine learning, the hinge loss function is used for training classifiers. For an output $t = \pm 1$ and a classifier score x, the hinge loss of the prediction $L(x)$ is defined as

$$L(x) = \max(0, 1 - t \cdot x) \tag{3.35}$$

if $t = -1$ and $x \geq 0$, then $L(x) = 1 + x > 0$; if $t = -1$ and $x < 0$, then $L(x) = \max(0, 1 - |x|)$; if $t = +1$, $x \geq 0$, $L(x) = \max(0, 1 - |x|)$; if $t = +1$, $x < 0$, $L(x) = 1 + x > 0$. Hence, $sgn(x) \cdot sgn(t) = -1$, $L(x) = 1 + x > 0$, where $sgn(\cdot)$ is the sign function and returns $+1$ or -1. Namely, $sgn(x) = +1$ if $x > 0$; $sgn(x) = -1$ if $x < 0$.

For $sgn(x) \cdot sgn(t) = +1$, $L(x) = \max(0, 1 - |x|)$. If $|x| > 1$, $1 - |x| < 0$, $L(x) = 0$; if $|x| < 1$, $1 - |x| > 0$, $L(x) = 1 - |x| > 0$. That means, $0 \leq x < 1$, $L(x) = 1 - x > 0$; if $-1 < x \leq 0$, $L(x) = 1 + x > 0$.

In summary, if $sgn(x) \cdot sgn(t) = -1$, then $L(x) = 1 + x > 0$; if $sgn(x) \cdot sgn(t) = +1$, $|x| < 1$, then $L(x) = 1 + x > 0$; if $sgn(x) \cdot sgn(t) = +1$, $|x| > 1$, $1 - |x| < 0$, then $L(x) = 0$.

In a loss function $y = f(x)$, we usually need to calculate the derivative $y' = \frac{df(x)}{dx}$, the chain rule is therefore applied if $x = s(t)$, then,

$$y' = \frac{df(x)}{dx} = \frac{df(x)}{dx} \cdot \frac{dx(t)}{dt}. \tag{3.36}$$

In most of time, we ensure the continuity of a function, but we cannot guarantee that the derivative exists. LSTM [46, 50–54] has been utilized to avoid this gradient varnishing or exploding problems. LSTM (Long short-term memory) is a typical RNN neural network,

$$\mathbf{f}_t = \sigma_g(\mathbf{W}_f \cdot \mathbf{x}_t + \mathbf{U}_f \cdot \mathbf{h}_{t-1} + \mathbf{b}_f) \tag{3.37}$$

$$\mathbf{i}_t = \sigma_g(\mathbf{W}_i \cdot \mathbf{x}_t + \mathbf{U}_i \cdot \mathbf{h}_{t-1} + \mathbf{b}_i) \tag{3.38}$$

$$\mathbf{o}_t = \sigma_g(\mathbf{W}_o \cdot \mathbf{x}_t + \mathbf{U}_o \cdot \mathbf{h}_{t-1} + \mathbf{b}_o) \tag{3.39}$$

$$\mathbf{c}_t = \mathbf{f}_t \cdot \mathbf{c}_{t-1} + \mathbf{i}_t \circ \sigma_c(\mathbf{W}_c \cdot \mathbf{x}_t + \mathbf{U}_c \cdot \mathbf{h}_{t-1} + \mathbf{b}_c) \tag{3.40}$$

$$\mathbf{h}_t = \mathbf{o}_t \circ \sigma_h(\mathbf{c}_t), \tag{3.41}$$

where $\mathbf{x}_t$ and $\mathbf{h}_t$ are the input and output vectors, respectively; $\mathbf{c}_0 = 0$, $\mathbf{h}_0 = 0$; $\mathbf{f}_t$, $\mathbf{i}_t$, and $\mathbf{o}_t$ are the activation vectors of forget, input, and output gates, respectively; $\mathbf{W}$, $\mathbf{U}$, $\mathbf{b}$ are the weight matrices and bias vector, respectively; $\mathbf{c}_t$ is the cell state vector; 'o' is the Hadamard product, i.e.,

$$\mathbf{A}_{m \times n} \cdot \mathbf{B}_{m \times n} = (a_{ij})_{m \times n} \cdot (b_{ij})_{m \times n} = (a_{ij} \cdot b_{ij})_{m \times n}, \tag{3.42}$$

where $\sigma_g(\cdot)$, $\sigma_c(\cdot)$, and $\sigma_h(\cdot)$ are the activation functions.

ConvLSTM (Convolutional LSTM) [54] used the spatiotemporal relationship,

$$\mathbf{f}_t = \sigma_g(\mathbf{W}_f \star \mathbf{x}_t + \mathbf{U}_f \star \mathbf{h}_{t-1} + \mathbf{V}_f \circ \mathbf{c}_{t-1} + \mathbf{b}_f) \tag{3.43}$$

$$\mathbf{i}_t = \sigma_g(\mathbf{W}_i \star \mathbf{x}_t + \mathbf{U}_i \star \mathbf{h}_{t-1} + \mathbf{V}_i \circ \mathbf{c}_{t-1} + b_i) \tag{3.44}$$

$$\mathbf{o}_t = \sigma_g(\mathbf{W}_o \star \mathbf{x}_t + \mathbf{U}_o \star \mathbf{h}_{t-1} + \mathbf{V}_o \circ \mathbf{c}_{t-1} + \mathbf{b}_o) \tag{3.45}$$

$$\mathbf{c}_t = \mathbf{f}_t \cdot \mathbf{c}_{t-1} + \mathbf{i}_t \circ \sigma_c(\mathbf{W}_c \star \mathbf{x}_t + \mathbf{U}_c \star \mathbf{h}_{t-1} + \mathbf{b}_c) \tag{3.46}$$

$$\mathbf{h}_t = \mathbf{o}_t \circ \sigma_h(\mathbf{c}_t), \tag{3.47}$$

where $\mathbf{x}_t$ and $\mathbf{h}_t$ are the input and output vectors, respectively, $\mathbf{c}_0 = 0$, $\mathbf{h}_0 = 0$; $\mathbf{f}_t$, $\mathbf{i}_t$, and $\mathbf{o}_t$ are the activation vectors of forget, input, and output gates, respectively; $\mathbf{W}$, $\mathbf{U}$, $\mathbf{V}$, $\mathbf{b}$ are the weight matrices and bias vector; $\mathbf{c}_t$ is the cell state vector; 'o' is the Hadamard product, '$\star$' is the convolution operator. $\sigma_g(\cdot)$, $\sigma_c(\cdot)$, and $\sigma_h(\cdot)$ are activation functions.

Furthermore, we have the peephole LSTM,

$$\mathbf{f}_t = \sigma_g(\mathbf{W}_f \cdot \mathbf{x}_t + \mathbf{U}_f \cdot \mathbf{c}_{t-1} + \mathbf{b}_f) \tag{3.48}$$

$$\mathbf{i}_t = \sigma_g(\mathbf{W}_i \cdot \mathbf{x}_t + \mathbf{U}_i \cdot \mathbf{c}_{t-1} + \mathbf{b}_i) \tag{3.49}$$

$$\mathbf{o}_t = \sigma_g(\mathbf{W}_o \cdot \mathbf{x}_t + \mathbf{U}_o \cdot \mathbf{c}_{t-1} + \mathbf{b}_o) \tag{3.50}$$

$$\mathbf{c}_t = \mathbf{f}_t \cdot \mathbf{c}_{t-1} + \mathbf{i}_t \circ \sigma_c(\mathbf{W}_c \cdot \mathbf{x}_t + \mathbf{U}_c \cdot \mathbf{h}_{t-1} + \mathbf{b}_c) \tag{3.51}$$

$$\mathbf{h}_t = \mathbf{o}_t \circ \sigma_h(\mathbf{c}_t), \tag{3.52}$$

where $\mathbf{x}_t$ and $\mathbf{h}_t$ are the input and output vectors, respectively, $\mathbf{c}_0 = 0$, $\mathbf{h}_0 = 0$; $\mathbf{f}_t$, $\mathbf{i}_t$, and $\mathbf{o}_t$ are the activation vectors of forget, input, and output gates, respectively; $\mathbf{W}$, $\mathbf{U}$, and $\mathbf{b}$ are the weight matrices and bias vector; $\mathbf{c}_t$ is the cell state vector; 'o' is the Hadamard product; $\sigma_g(\cdot)$, $\sigma_c(\cdot)$, and $\sigma_h(\cdot)$ are the activation functions.

Meanwhile, we have the FRU (Fully gated unit), which has the following steps:

Initially, $t = 0$, $\mathbf{h}_0 = 0$,

$$\mathbf{z}_t = \sigma_g(\mathbf{W}_z \cdot \mathbf{x}_t + \mathbf{U}_z \cdot \mathbf{h}_{t-1} + \mathbf{b}_z) \quad (\textit{update gate}) \tag{3.53}$$

$$\mathbf{r}_t = \sigma_g(\mathbf{W}_r \cdot \mathbf{x}_t + \mathbf{U}_r \cdot \mathbf{h}_{t-1} + \mathbf{b}_r) \quad (\textit{reset gate}) \tag{3.54}$$

$$\tilde{\mathbf{h}}_t = \sigma_h(\mathbf{W}_h \cdot x_t + \mathbf{U}_h(r_t \circ \mathbf{h}_{t-1}) + \mathbf{b}_h) \quad (\textit{new memory}) \tag{3.55}$$

$$\mathbf{h}_t = (1 - \mathbf{z}_t) \circ \mathbf{h}_{t-1} + \mathbf{z}_t \circ \tilde{\mathbf{h}}_t \quad (\textit{hidden state}), \tag{3.56}$$

where $\mathbf{x}_t$ and $\mathbf{h}_t$ are the input and output vectors, respectively, $\mathbf{W}$, $\mathbf{U}$, and $\mathbf{b}$ are the weight matrices and vector; 'o' is the Hadamard product. $\sigma_g(\cdot)$ and $\sigma_h(\cdot)$ are the sigmoid function and tanh function, respectively.

For simplifying the problem, the Minimal Gated Unit (MGU) has the following steps:

Initially, $t = 0$, $\mathbf{h}_0 = 0$,

$$\mathbf{f}_t = \sigma_g(\mathbf{W}_f \cdot \mathbf{x}_t + \mathbf{U}_f \cdot \mathbf{h}_{t-1} + \mathbf{b}_f) \tag{3.57}$$

$$\mathbf{h}_t = \mathbf{f}_t \circ \mathbf{h}_{t-1} + (1 - \mathbf{f}_t) \circ \sigma_h(\mathbf{W}_h \cdot x_t + \mathbf{U}_h(\mathbf{f}_t \circ \mathbf{h}_{t-1}) + \mathbf{b}_h), \tag{3.58}$$

where $\mathbf{x}_t$ and $\mathbf{h}_t$ are the input and output vectors, respectively, $\mathbf{f}_t$ is the forget vector; $\mathbf{W}$, $\mathbf{U}$, and $\mathbf{b}$ are the weight matrices and vector; 'o' is the Hadamard product. $\sigma_g(\cdot)$ and $\sigma_h(\cdot)$ are the sigmoid functions, tanh is the activation function.

3.3.2   Time Series Analysis

With time changes, the states will be altered. Time series analysis [55] is applied to deal with these state-changing issues. For example, time series analysis has been applied to water quality control or air quality assessment with time changes. We need to find out the patterns behind time series analysis after pattern observations.

Observation is one of the steps of artificial intelligence (AI) besides learning, presentation, and inference or reasoning. AI can predict what will happen from what have already happened, we also call it forecasting or prediction [56, 57].

There are two main goals of time series analysis: Identifying the nature of the phenomenon represented by using a sequence of observations; forecasting is to predict future values of the time series variable.

Most of time series patterns is described in terms of a basic class of components: Trend analysis (smoothing, fitting a function, etc.), analysis of seasonality (autocorrelation correlogram, examining correlograms, partial autocorrelations, removing serial dependency, etc.)

In time series analysis, we use the concept seasonality, that means the patterns usually follow the seasons annually. We can use the changes of data in time series to analyse the patterns.

In time analysis, for a sequence $x(1), x(2), \ldots, x(t), \ldots$, there are

- Mean:
$$\mu_t = \mathrm{E}(X(t)) \tag{3.59}$$

- Variance:
$$\sigma^2 = \mathrm{Var}(X(t)) \tag{3.60}$$

- Autocovariance:
$$\gamma(t_1, t_2) = \mathrm{E}\{[X(t_1) - \mu(t_1)][X(t_2) - \mu(t_2)]\} \tag{3.61}$$

- Autocovariance lag:

$$\gamma(\tau) = \mathrm{E}\{[X(t) - \mu][X(t + \tau) - \mu]\} \tag{3.62}$$

- Autocovariance function:

$$\rho(\tau) = \gamma(\tau)/\gamma(0) \tag{3.63}$$

$$\rho(\tau) = \rho(-\tau), |\rho(\tau)| < 1 \tag{3.64}$$

- Random walk:

$$X_t = X_{t-1} + Z_t \tag{3.65}$$

- MA(q) process:

$$X_t = \beta_0 Z_t + \beta_1 Z_{t-1} + \cdots + \beta_q Z_{t-q} \tag{3.66}$$

- AR(p) process:

$$X_t = \alpha_1 X_{t-1} + \cdots + \alpha_p X_{t-p} + Z_t. \tag{3.67}$$

- Mixed autoregressive moving average model (ARMA)(p, q):

$$X_t = \alpha_1 X_{t-1} + \cdots + \alpha_p X_{t-p} + Z_t + \beta_0 Z_t + \cdots + \beta_q Z_{t-q} \tag{3.68}$$

- Exponential smoothing:

$$S(x_t) = \sum_{j=0}^{\infty} \alpha(1 - \alpha)^j x_{t-j} \tag{3.69}$$

- Residual:

$$R(x_t) = x_t - S(x_t) \tag{3.70}$$

$$a_j = \alpha(1 - \alpha)^j. \tag{3.71}$$

- Convolution:

$$\{c_k\} = \{a_r\} \star \{b_j\} \Leftrightarrow c_k = \sum_r a_r \cdot b_{k-r} \tag{3.72}$$

$$z_t = \sum_k c_k \cdot x_{t+k} = \sum_j b_j \cdot y_{t+j}. \tag{3.73}$$

- Additive seasonality:

$$X_t = m_t + S_t + \varepsilon_t. \tag{3.74}$$

- Multiplicative seasonality:

$$X_t = m_t \cdot S_t + \varepsilon_t. \tag{3.75}$$

In time series analysis, we usually need to remove the noises from the collected data. Usually, a kernel for convolution operation is needed [58]. The convolutional operation will reduce the noises and make the signals smooth.

In time analysis, we usually need to use spectrum analysis. Fourier series [58] has been applied to signal decomposition and multiresolution analysis based on trigonometric function function because sine and cosine functions construct an orthogonal function system.

$$f(t) \approx \frac{a_0}{2} + \sum_{r=1}^{k}(a_r \cos rt + b_r \sin rt) \tag{3.76}$$

where

$$a_0 = \frac{1}{\pi}\int_{-\pi}^{\pi} f(t)dt \tag{3.77}$$

$$a_r = \frac{1}{\pi}\int_{-\pi}^{\pi} f(t)\cos rt\,dt \tag{3.78}$$

$$b_r = \frac{1}{\pi}\int_{-\pi}^{\pi} f(t)\sin rt\,dt. \tag{3.79}$$

Kalman filter is a linear model for object tracking or signal filtering. Kalman filter has an updating function and a predicting function. Dynamical updates have been applied to update the parameters for the purpose of prediction and correction.

- Prediction equations:

$$\theta_{t|t-1} = G_t\theta_{t-1} \tag{3.80}$$

and

$$P_{t|t-1} = G_t P_{t-1} G_t^{\tau} + w_t. \tag{3.81}$$

- Updating equations:

$$\theta_t = \theta_{t|t-1} + K_t e_t \tag{3.82}$$

and

$$P_t = P_{t|t-1} - K_t h_t^{\tau} P_{t|t-1}, \tag{3.83}$$

where

$$K_t = P_{t|t-1}h_t/[h_t^{\tau} P_{t|t-1}h_t + \sigma_n^2]. \tag{3.84}$$

For a nonlinear system in time series analysis, we have

- Nonlinear autoregressive model (NLAR):

$$X_t = f(X_{t-1}, X_{t-2}, \dots, X_{t-p}) + Z_t. \tag{3.85}$$

- Threshold autoregressive model:

$$X_t = \begin{cases} \alpha_1 X_{t-1} + Z_t \text{ if } X_{t-1} < r \\ \alpha_2 X_{t-1} + Z_t \text{ if } X_{t-1} \geq r \\ \quad . \end{cases} \tag{3.86}$$

- Artificial neural networks:

$$y = \phi_0[\sum_j w'_j \phi_h(\sum_i w_{ij}x_i) + w'_0],\qquad (3.87)$$

where y is the output, $v_j = \sum_i w_{ij}x_i$, ϕ_h is the activation function.

LSTM has been applied to time series analysis for forecasting, especially MAT-LAB has provided a program for predicting or forecasting; meanwhile, LSTM can provide RMS errors for the prediction. The predictions are much accurate when updating the network state with the observed values instead of the predicted values.

In order to forecast the values of future time step of a sequence, we train a regression LSTM network, where the responses are the training sequences with values shifted by one time step. That is, at each time step of the input sequence, the LSTM network learns to predict the value of the next time step.

If we access to the actual values of time steps between predictions, then we can update the network state with the observed values. For each prediction, we predict the next time step by using the observed value of the previous time step. We calculate the root-mean-square error (RMSE).

Abnormal detection or anomaly detection is a typical application of time series analysis [59].

3.4   Functional Spaces

In deep learning, we need to calculate the loss function; actually, it is a kind of distance. In this section, we will introduce how to measure the distance in functional spaces.

3.4.1   Metric Space

Metric spaces [60] are thought as very basic ones where the ideas of convergence and continuity exist. Distance or metric is a measure of how close two elements are to each other [57].

A distance (or metric) on a metric space $\mathbf{X}$ is a function: $d : \mathbf{X}^2 \to \mathbf{Y} \subset \mathscr{R}^+$; $(x, y) \mapsto d(x, y) = \|x - y\|, x, y, z \in \mathbf{X}$.

- Triangle inequality: $d(x, y) \le d(x, z) + d(z, y)$
- Symmetry: $d(x, y) = d(y, x)$
- Equality: $d(x, y) = 0 \Leftrightarrow x = y$.

Therefore,

- $d(x, y) \geq |d(x, z) - d(z, y)|$
- $x_1, x_2, \ldots, x_n \in \mathbf{X}, d(x_1, x_n) \leq \sum_{i=1}^{n-1} d(x_i, x_{i+1})$.

A sequence (x_n) in a metric space $\mathbf{X}$ converges to a limit x, denoted as $\lim_{n \to \infty} x_n = x$, when $\forall \varepsilon > 0, \exists N, n \gg N \Rightarrow x_n \in B_\varepsilon(x)$.

A function $f : \mathbf{X} \to \mathbf{Y}$ between metric spaces is continuous when it preserves convergence, $x_n \to x \in \mathbf{X} \Rightarrow f(x_n) \to f(x) \in \mathbf{Y}$. If a function f is continuous, it is invertible and its inverse $f^{-1}(x)$ is continuous.

A Cauchy sequence is such one that $d(x_n, x_m) \to 0$ as $n, m \to \infty$; namely, $\forall \varepsilon > 0, \exists N$, if $n, m \geq N$, then $d(x_n, x_m) < \varepsilon$.

A sequence $x_1, x_2, \cdots, x_n, \ldots$ is Cauchy sequence, if and only if, every subsequence is asymptotic to this sequence.

A uniformly continuous function maps any Cauchy sequence to a Cauchy sequence. A function is *Uniformly continuous*, which refers to δ is independent on x_i.

A function $f : \mathbf{X} \to \mathbf{Y}$ is a *Lipschitz map* ,when $\exists c > 0, \forall x, x' \in \mathbf{X}, d_{\mathbf{Y}}(f(x), f(x')) \leq c \cdot d_{\mathbf{X}}(x, x')$.

A metric space is complete when every Cauchy sequence converges, e.g., the real space is complete.

A metric space is separable when it contains a countable dense subset, $\exists \mathbf{A} \subseteq \mathbf{X}$, $\mathbf{A}$ is countable and $\bar{\mathbf{A}} = \mathbf{X}$.

A set $\mathbf{B}$ is bounded when the distance between any two points in the set has an upper bound. $\exists r > 0, \forall x, y \in \mathbf{B}, d(x, y) \leq r$.

The least such upper bound is called the diameter of the set: $diam \ \mathbf{B} := \sup_{x, y \in \mathbf{B}} d(x, y)$.

A set $\mathbf{K}$ is compact given any cover of balls, there is a finite subcollection of them that still cover the set (a subcover) $K \subseteq \bigcup_i B_{\varepsilon_i}(a_i) \Rightarrow K \subseteq \bigcup_{n=1}^N B_{\varepsilon_{i_n}}(a_{i_n})$.

3.5   Vector Space

Distance is a scalar value. Given two n-dimensional vectors $\mathbf{x} = (x_1, x_2, \ldots, x_n)^\tau$ and $\mathbf{y} = (y_1, y_2, \ldots, y_n)^\tau$, Euclidean distance is

$$d = \sqrt{\sum_{i=1}^n (x_i - y_i)^2}. \tag{3.88}$$

Manhattan distance is

$$d = \sum_{i=1}^n (|x_i - y_i|). \tag{3.89}$$

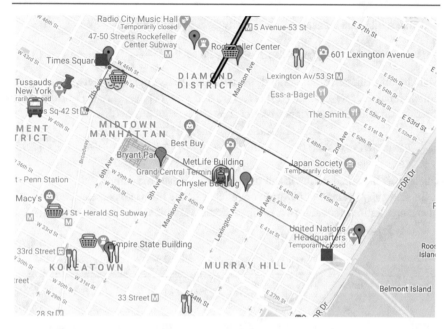

Fig. 3.1 The Google Map at Manhattan downtown in New York City. The distance between United Nations Headquarter and Times Square is roughly same no matter which way will be taken in the city (red and blue)

Manhattan is a region of New York City, New York, USA. An example of distance is shown in Fig. 3.1 for the distance from Times Square to the United Nations Headquarter in Manhattan downtown area.

Chebyshev distance is

$$d = \max_{i=1}^{n}(|x_i - y_i|). \tag{3.90}$$

Minkowski distance is a metric which is a generalization of both the Euclidean distance and the Manhattan distance.

$$d = \lim_{n \to +\infty}\left(\sum_{i=1}^{n}(x_i - y_i)^p\right)^{\frac{1}{p}} = \max_{i=1}^{n}(|x_i - y_i|) \tag{3.91}$$

and

$$d = \lim_{n \to -\infty}\left(\sum_{i=1}^{n}(x_i - y_i)^p\right)^{\frac{1}{p}} = \min_{i=1}^{n}(|x_i - y_i|). \tag{3.92}$$

Given sets **A** and **B**, Jaccard index, also known as Intersection over Union (IoU), or the Jaccard similarity coefficient is

$$J(\mathbf{A}, \mathbf{B}) = \frac{\mathbf{A} \cap \mathbf{B}}{\mathbf{A} \cup \mathbf{B}}. \tag{3.93}$$

Jaccard distance measures dissimilarity between sample sets

$$J_d(\mathbf{A}, \mathbf{B}) = 1 - J(\mathbf{A}, \mathbf{B}) = 1 - \frac{\mathbf{A} \cap \mathbf{B}}{\mathbf{A} \cup \mathbf{B}}. \tag{3.94}$$

Mahalanobis distance is defined as a dissimilarity measure between two n-dimensional vectors $\mathbf{x} = (x_1, x_2, \ldots, x_n)^\tau$ and $\mathbf{y} = (y_1, y_2, \ldots, y_n)^\tau$ of the same distribution with the covariance matrix Σ

$$d(\mathbf{x}, \mathbf{y}) = \sqrt{(\mathbf{x} - \mathbf{y})^\tau \Sigma^{-1} (\mathbf{x} - \mathbf{y})} = \sqrt{\frac{\sum_{i=1}^n (x_i - y_i)^2}{\sigma_i^2}}, \tag{3.95}$$

where σ_i is the standard deviation of the x_i and y_i over the sample set.

In statistics, Pearson correlation coefficient or the bivariate correlation measures linear correlation between two n-dimensional variables $\mathbf{x} = (x_1, x_2, \ldots, x_n)$ and $\mathbf{y} = (y_1, y_2, \ldots, y_n)$.

$$d(\mathbf{x}, \mathbf{y}) = \frac{cov(\mathbf{x}, \mathbf{y})}{\sigma_\mathbf{x} \sigma_\mathbf{y}} = \frac{\mathscr{E}(\mathbf{x} - \bar{\mathbf{x}})(\mathbf{y} - \bar{\mathbf{y}})}{\sigma_\mathbf{x} \sigma_\mathbf{y}} = \frac{\sum_{i=1}^n (x_i - \bar{x})(y_i - \bar{y})}{\sqrt{\sum_{i=1}^n (x_i - \bar{x})^2} \sqrt{\sum_{i=1}^n (y_i - \bar{y})^2}}, \tag{3.96}$$

where $\bar{x} = \frac{\sum_{i=1}^n x_i}{n}$ and $\bar{y} = \frac{\sum_{i=1}^n y_i}{n}$ are the means, $cov(\mathbf{x}, \mathbf{y})$ is the covariance, $\sigma_\mathbf{x}$ is the standard deviation of $\mathbf{x}$, $\sigma_\mathbf{y}$ is the standard deviation of $\mathbf{y}$, $\mathscr{E}$ is the expectation.

A vector space [60] $\mathbf{V}$ over a field $\mathbf{F}$ is a set on which an operation of vector addition is defined $+ : \mathbf{V}^2 \to \mathbf{V}^2$ satisfying associativity, commutativity, zero and inverse axioms:

- For every $x, y, z \in \mathbf{V}$,
  $x + (y + z) = (x + y) + z; x + y = y + x;$
  $0 + x = x; x + (-x) = 0;$
- An operation of scalar multiplication $\mathbf{F} \times \mathbf{V} \to \mathbf{V}$ that satisfies the respective distributive laws.
- For every $\lambda, \mu \in \mathbf{F}$,
  $\lambda(x + y) = \lambda x + \lambda y; (\lambda + \mu)x = \lambda x + \mu x;$
  $(\lambda \mu)x = \lambda(\mu x); 1x = x.$

Every vector has a base. For an n-dimensional vector $\mathbf{V} = (v_1, v_2, \ldots, v_n)^\tau$, we have the corresponding p-norm $\|\mathbf{V}\|_p = (\sum_{i=1}^n v_i^p)^{\frac{1}{p}}$, $p = 0, 1, 2, \ldots, \infty$. If $p = 1$, then it is 1-norm $\|\mathbf{V}\|_1 = (|\sum_{i=1}^n v_i|)$; if $p = 0$, then it is zero norm $\|\cdot\|_0 = min_i(|v_i|)$; if $p = \infty$, then it is infinity norm or maximum norm $\|\cdot\|_\infty = max_i(|v_i|)$.

3.5.1 Normed Space

A norm on a real vector space $\mathbf{X}$ is a function: $\mathbf{X} \to \mathscr{R}$, $\mathbf{u} \mapsto \|\mathbf{u}\|$,

- $\forall \mathbf{u} \in \mathbf{X}$, $\|\mathbf{u}\| \geq 0$.
- $\forall \mathbf{u} \in \mathbf{X}$, $\alpha \in \mathbf{R}$, $\|\alpha \mathbf{u}\| = |\alpha| \|\mathbf{u}\|$.
- $\forall \mathbf{u}, \mathbf{v} \in \mathbf{X}$, $\|\mathbf{u} + \mathbf{v}\| \leq \|\mathbf{u}\| + \|\mathbf{v}\|$. (Minkowski's inequality)

A normed space $\mathbf{X}$ is a vector space over $\mathbf{F} = \mathbf{R}$ or $\mathbf{C}$ with a function called the norm $\| \cdot \| : \mathbf{X} \mapsto \mathscr{R}$ such that for any $x, y \in \mathbf{X}$, $\lambda \in \mathbf{F}$, $\|x + y\| \leq \|x\| + \|y\|$, $\|\lambda x\| = \lambda \|x\|$, $\|x\| = 0 \Leftrightarrow x = 0$. Thus,

$$\|x - y\| \geq \|x\| - \|y\| \tag{3.97}$$

and

$$\|x_1 + x_2 + \cdots + x_n\| \leq \|x_1\| + \|x_2\| + \cdots + \|x_n\|. \tag{3.98}$$

Given

$$\|(a_n)\|_2 = \sqrt{\sum_{n=0}^{\infty} \|a_n\|^2} \tag{3.99}$$

and

$$\|(b_n)\|_2 = \sqrt{\sum_{n=0}^{\infty} \|b_n\|^2}. \tag{3.100}$$

Cauchy's inequality is

$$\left| \sum_{n=0}^{\infty} a_n b_n \right| \leq \|(a_n)\|_2 \|(b_n)\|_2 \tag{3.101}$$

and

$$\sqrt{\sum_{n=0}^{\infty} \|a_n + b_n\|^2} \leq \|(a_n)\|_2 + \|(b_n)\|_2. \tag{3.102}$$

- Vector addition, scalar multiplication, and the norm are continuous.
- When (x_n) and (y_n) converge, $(x_n + y_n)$, (λx_n) and $(\|(x_n)\|)$ converge, namely,

$$\lim_{n \to \infty} (x_n + y_n) = \lim_{n \to \infty} x_n + \lim_{n \to \infty} y_n \tag{3.103}$$

$$\lim_{n \to \infty} (\lambda x_n) = \lambda \lim_{n \to \infty} x_n \tag{3.104}$$

$$\lim_{n \to \infty} \|x_n\| = \| \lim_{n \to \infty} x_n \|. \tag{3.105}$$

3.5.2 Hilbert Space

Hilbert spaces are Banach spaces with a norm derived from a scalar product [60]. A scalar product on the (real) vector space $\mathbf{X}$ is a function: $(u, v) \in \mathbf{X} \times \mathbf{X} \to \mathbf{R}$, $(u, v) \mapsto (u|v)$.

- $\forall \mathbf{u} \in \mathbf{X}, (\mathbf{u}|\mathbf{u}) \geq 0$.
- $\forall \mathbf{u}, \mathbf{v}, \mathbf{w} \in \mathbf{X}, \alpha, \beta \in \mathbf{R}, (\alpha\mathbf{u} + \beta\mathbf{v}|\mathbf{w}) = \alpha(\mathbf{u}|\mathbf{w}) + \beta(\mathbf{v}|\mathbf{w})$. $\|\mathbf{u}\| = \sqrt{(u|u)}$.

A Hilbert space is an inner product space which is complete as a metric space. An inner product on a vector space is a positive-definite sesquilinear form: $<, >$: $\mathbf{X} \times \mathbf{X} :\mapsto \mathbf{F}$. For $x, y, z \in \mathbf{X}, \lambda \in \mathbf{F}$,

- $< x, y + z > = < x, y > + < x, z >$; $< x, \lambda y > = \lambda < x, y >$; $< x, y > = \overline{< y, x >}$; $< x, x > \geq 0$; $< x, x > = 0 \Rightarrow x = 0$.
- Cauchy–Schwarz inequality:

$$| < x, y > | \leq \|x\|\|y\| \tag{3.106}$$

- Pythagoras' theorem:

$$\|x + y\| \leq \|x\| + \|y\|. \tag{3.107}$$

- The inner product is continuous:

$$\lim_{n \to \infty} < x_n, y_n > = < \lim_{n \to \infty} x_n, \lim_{n \to \infty} y_n > . \tag{3.108}$$

- A norm is induced from an inner product, if and only if, it satisfies, for all vectors $x, y \in \mathscr{R}$,

$$\|x + y\|^2 + \|x - y\|^2 = 2(\|x\|^2 + \|y\|^2). \tag{3.109}$$

The orthogonal spaces of subsets $\mathbf{A} \subseteq \mathbf{X}$,

$$\mathbf{A}^\perp := \{\forall x \in X, < x, a > = 0, \forall a \in \mathbf{A}\} \tag{3.110}$$

satisfy, $\mathbf{A} \cap \mathbf{A}^\perp \subseteq 0$; $\mathbf{A} \subseteq \mathbf{B} \Leftrightarrow \mathbf{B}^\perp \subseteq \mathbf{A}^\perp$; $\mathbf{A} \subseteq \mathbf{A}^{\perp\perp}$; $\mathbf{A}^\perp$ is a closed subspace of $\mathbf{X}$.

If $\mathbf{M}$ is a closed convex subset of a Hilbert space $\mathbf{H}$, then any point in $\mathbf{H}$ has a unique point in $\mathbf{M}$ which is closest to it by using the least squares approximation.

An orthonormal basis of a Hilbert space $\mathbf{H}$ is a set of orthonormal vectors $\mathbf{E}$ whose span is dense: $\forall e_i, e_j \in \mathbf{E}, < e_i, e_j > = \delta_{ij}$.

Parseval's identity (Fourier Series): If $x = \sum_{i=1}^n \alpha_i e_i$, $y = \sum \beta_i e_i$, $\{e_i\}$ is orthonormal, then $x, y \in \mathbf{H}$, $x = \sum \alpha_i e_i$, $< x, y > = \sum < x, e_i > < e_i, y >$, $\sum | < x, e_i > |^2 = \|x\|^2$.

Bessel's inequality: $x = \sum \alpha_i e_i$, $\sum | < x, e_i > |^2 \leq \|x\|^2$.

An instance in Hilbert space is Fourier transform, which refers to both the frequency domain representation and the mathematical operation that associates the frequency domain representation to a function of time [61]. The Fourier transform is an extension of the Fourier series. If we increase the length of the interval in the

Fourier series, then the Fourier coefficients begin to resemble the Fourier transform. The Fourier transform of a function of time is a complex-valued function of frequency, whose magnitude (absolute value) represents the amount of that frequency present in the original function, and whose argument is the phase offset of the basic sinusoid in that frequency.

The Fourier transform of a function f is traditionally denoted $\hat{f}$, the Fourier transform of an integrable function $f : \mathscr{R} \to \mathscr{C}$,

$$\hat{f}(\xi) = \int_{-\infty}^{\infty} f(x)e^{-2\pi i x \xi} dx, \tag{3.111}$$

for any real number ξ.

When the independent variable x represents time, the transform variable ξ represents frequency, f is determined by $\hat{f}$ via the inverse transform

$$f(\xi) = \int_{-\infty}^{\infty} \hat{f}(\xi)e^{2\pi i x \xi} d\xi \tag{3.112}$$

for any real number x.

2D discrete Fourier transform (DFT) maps a scalar image I from spatial domain into a complex-valued Fourier transform $\mathbf{I}$ in frequency domain.

$$\mathbf{I}(u, v) = \frac{1}{W \cdot H} \sum_{x=0}^{W-1} \sum_{y=0}^{H-1} I(x, y)exp\left[-i2\pi \left(\frac{x \cdot u}{W} + \frac{y \cdot v}{H}\right)\right], \tag{3.113}$$

where $u = 0, 1, \ldots, W-1$ and $v = 0, 1, \ldots, H-1, i = \sqrt{-1}$ as the imaginary unit of complex numbers, W and H are the width and height of the image, respectively.

The inverse 2D DFT maps a Fourier transform $\mathbf{I}$ in frequency domain back into the spatial domain.

$$I(x, y) = \sum_{u=0}^{W-1} \sum_{v=0}^{H-1} \mathbf{I}(u, v)exp\left[i2\pi \left(\frac{x \cdot u}{W} + \frac{y \cdot v}{H}\right)\right]. \tag{3.114}$$

The discrete Fourier transform (DFT) for an image I satisfies Parseval's theorem

$$\frac{1}{|\Omega|} \sum_{\Omega} |I(x, y)|^2 = \sum_{\Omega} |\mathbf{I}(u, v)|^2, \tag{3.115}$$

where $\Omega = [1, W] \times [1, H]$.

The screen shots are shown in Fig. 3.2. Figure 3.2a displays 1D Fourier transform, Fig. 3.2b indicates 2D Fourier transform.

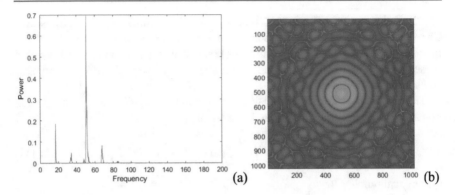

Fig. 3.2 An example of Fourier transform by using MATLAB, **a** 1D, **b** 2D

3.6 Questions

Question 1. Can the YOLOv2 cope with small target detection and classification such as cell ?

Question 2. Please explain the relationships between deep learning concepts: RNN, LSTM, GRU, FGU.

Question 3. How to merge or fuse different networks such as U-net and YOLOv2?

Question 4. In deep learning, how to select a proper algorithm for object detection?

Question 5. What is the difference between HMM and RNN?

Question 6. How to select the loss function in deep learning?

Question 7. For time series analysis, what are the advantages of deep learning methods?

Question 8. How to understand cost functions of artificial neural networks are a kind of metric or distance in functional spaces?

Question 9 What is the relationship between norm and regularization in deep learning from the viewpoint of functional analysis?

Question 10 How to understand the relationship between Fourier transform and Hilbert space?

References

1. Krizhevsky A, Sutskever I, Hinton G (2017) ImageNet classification with deep convolutional neural networks. Commun ACM 60(6):84–90
2. Krizhevsky A, Sutskever I, Hinton GE (2012) ImageNet classification with deep convolutional neural networks. In: Advances in Neural Information Processing Systems, pp 1097–1105
3. Rastegari M, Ordonez V, Redmon J, Farhadi A (2016) XNOR-Net: ImageNet classification using binary convolutional neural networks. In: European Conference on Computer Vision, pp 525–542. Springer, Berlin

4. Russakovsky O, Deng J, Su H, Krause J, Satheesh S, Ma S, Berg AC (2015) ImageNet large scale visual recognition challenge. Int J Comput Vis 115(3):211–252
5. LeCun Y, Bengio Y, Hinton G (2015) Deep learning. Nature 521:436–444
6. Vapnik VN (1995) The nature of statistical learning theory. Springer, Berlin
7. Zanaty EA (2012) Support vector machines (SVMs) versus multilayer perception (MLP) in data classification. Egypt Inf J 13(3):177–183
8. LeCun Y, Bengio Y (1995) Convolutional networks for images, speech, and time series. Handbook Brain Theory Neural Netw 3361(10):1995
9. Aizenberg NN, Aizenberg IN, Krivosheev GA (1996) CNN based on universal binary neurons: learning algorithm with error-correction and application to impulsive-noise filtering on grayscale images. In: IEEE international workshop on cellular neural networks and their applications, pp 309–314
10. Rekeczky C, Tahy A, Vegh Z, Roska T (1999) CNN-based spatio-temporal nonlinear filtering and endocardial boundary detection in echocardiography. Int J Circuit Theory Appl 27(1):171–207
11. Sahiner B, Chan HP, Petrick N, Wei D, Helvie MA, Adler DD, Goodsitt MM (1996) Classification of mass and normal breast tissue: a convolution neural network classifier with spatial domain and texture images. IEEE Trans Med Imag 15(5):598–610
12. Hubel DH, Wiesel TN (1962) Receptive fields, binocular interaction and functional architecture in the cat's visual cortex. J Physiol 160(1):106–154
13. Lee CY, Gallagher PW, Tu Z (2016) Generalizing pooling functions in convolutional neural networks: mixed, gated, and tree. In: Artificial intelligence and statistics, pp 464–472
14. Giusti A, Ciresan DC, Masci J, Gambardella LM, Schmidhuber J (2013) Fast image scanning with deep max-pooling convolutional neural networks. In: IEEE International conference on image processing, pp 4034–4038
15. Heikkila M, Pietikainen M (2006) A texture-based method for modeling the background and detecting moving objects. IEEE Trans Pattern Anal Mach Intell 28(4):657–662
16. He K, Zhang X, Ren S, Sun J (2014) Spatial pyramid pooling in deep convolutional networks for visual recognition. In: European conference on computer vision, pp 346–361. Springer, Berlin
17. Merrienboer B, Bahdanau D, Dumoulin V, Serdyuk D, Warde-Farley Murtagh, F (1991) Multilayer perceptrons for classification and regression. Neurocomputing 2(5–6):183–197
18. Taud H, Mas JF (2018) Multilayer perceptron (MLP). In: Geomatic approaches for modelling land change scenarios, pp 451–455. Springer, Berlin
19. Dai J, Li Y, He K, Sun J (2016) R-FCN: object detection via region-based fully convolutional networks. In: Advances in neural information processing systems, pp 379–387
20. Girshick R, Donahue J, Darrell T, Malik J (2016) Region-based convolutional networks for accurate object detection and segmentation. IEEE Trans Pattern Anal Mach Intell 38(1):142–158
21. Gkioxari G, Girshick R, Malik J (2015) Contextual action recognition with R-CNN. In: IEEE ICCV, pp 1080–1088
22. Girshick R (2015) Fast R-CNN. In: IEEE International conference on computer vision, pp 1440–1448
23. Gu Q, Yang J, Yan WQ, Li Y, Klette R (2017) Local Fast R-CNN flow for object-centric event recognition in complex traffic scenes. In: Pacific-rim symposium on image and video technology, pp 439–452
24. Kivinen J, Warmuth MK (1998) Relative loss bounds for multidimensional regression problems. In: Advances in neural information processing systems, pp 287–293
25. Rriedman J, Hastie T, Tibshirani R (2000) Additive logistic regression: a statistical view of boosting. Ann Stat 38(2):337–374
26. Ren S, He K, Girshick R, Sun J (2015) Faster R-CNN: towards real-time object detection with region proposal networks. In: Advances in neural information processing systems, pp 91–99

27. Ren Y, Zhu C, Xiao S (2018) Object detection based on fast/faster RCNN employing fully convolutional architectures. Math Probl Eng
28. Dunne RA, Campbell NA (1997) On the pairing of the softmax activation and cross-entropy penalty functions and the derivation of the softmax activation function. In: Australian Conference on the Neural Networks, Melbourne, vol 181, pp 185
29. Takeda F, Omatu S (1995) A neuro-paper currency recognition method using optimized masks by genetic algorithm. In: IEEE International conference on systems, man and cybernetics, vol 5, pp 4367–4371
30. Redmon J, Divvala S, Girshick R, Farhadi A (2016) You only look once: unified, real-time object detection. In: IEEE CVPR, pp 779–788
31. Redmon J, Farhadi A (2017) YOLO9000: better, faster, stronger. In: IEEE CVPR, pp 6517–6525
32. Liu W, Anguelov D, Erhan D, Szegedy C, Reed S, Fu CY, Berg AC (2016) SSD: single shot multibox detector. In: European conference on computer vision, pp 21–37
33. Cao G, Xie X, Yang W, Liao Q, Shi G, Wu J (2018) Feature-fused SSD: fast detection for small objects. In: International conference on graphic and image processing (ICGIP), vol 10615
34. Hager GD, Dewan M, Stewart CV (2004) Multiple kernel tracking with SSD, In: CVPR
35. Jeong J, Park H, Kwak N (2017) Enhancement of SSD by concatenating feature maps for object detection. In: BMVC'17
36. Huang G, Liu Z, Weinberger KQ, van der Maaten L (2017) Densely connected convolutional networks. In: IEEE CVPR, vol 1, issue 2, p 3
37. Hochreiter S, Schmidhuber J (1997) Long short-term memory. Neural Comput 9(8):1735–1780
38. Rabiner L, Juang B (1986) An introduction to hidden Markov models. IEEE ASSP (magazine) 3(1):4–16
39. Hassanpour H, Farahabadi PM (2009) Using hidden Markov models for paper currency recognition. Expert Syst Appl 36(6):10105–10111
40. Chatzis SP, Kosmopoulos DI (2011) A variational Bayesian methodology for hidden Markov models utilizing Student's-t mixtures. Pattern Recogn 44(2):295–306
41. Toselli AH, Vidal E, Romero V, Frinken V (2016) HMM word graph based keyword spotting in handwritten document images. Inf Sci 370:497–518
42. Gal Y, Ghahramani Z (2016) A theoretically grounded application of dropout in recurrent neural networks. In: Advances in neural information processing systems, pp 1019–1027
43. Mikolov T, Karafiat M, Burget L, Cernocky J, Khudanpur S (2010) Recurrent neural network based language model. In: Interspeech, vol 2, p 3
44. Martens J, Sutskever I (2011) Learning recurrent neural networks with Hessian-free optimization. In: International conference on machine learning, Bellevue
45. Gers FA, Schmidhuber J (2000) Recurrent nets that time and count. In: Proceedings of the IEEE-INNS-ENNS international joint conference on neural networks, vol 3, pp 189–194
46. Gers FA, Schraudolph NN, Schmidhuber J (2002) Learning precise timing with LSTM recurrent networks. J Mach Learn Res 3:115–143
47. Basu AP, Ebrahimi N (1991) Bayesian approach to life testing and reliability estimation using asymmetric loss function. J Stat Plann Inf 29(1–2):21–31
48. Liu W, Wen Y, Yu Z, Yang M (2016) Large-margin softmax loss for convolutional neural networks. In: ICML, pp 507–516
49. Zhang K, Zhang D, Jing C, Li J, Yang L (2017) Scalable softmax loss for face verification. In: International conference on systems and informatics, pp 491–496
50. Fu R, Zhang Z, Li L (2016) Using LSTM and GRU neural network methods for traffic flow prediction. In: Youth academic annual conference of Chinese association of autDomation (YAC)
51. Gers FA, Schmidhuber J, Cummins F (2000) Learning to forget: continual prediction with LSTM. Neural Comput 12(10):2451–2471
52. Gers FA, Schmidhuber E (2001) LSTM recurrent networks learn simple context-free and context-sensitive languages. IEEE Trans Neural Netw 12(6):1333–1340

53. Wang MS, Song L, Yang XK, Luo CF (2016) A parallel-fusion RNN-LSTM architecture for image caption generation. In: International conference on image processing, pp 4448–4452
54. Xingjian SHI, Chen Z, Wang H, Yeung DY, Wong WK, Woo WC (2015) Convolutional LSTM network: a machine learning approach for precipitation nowcasting. In: Advances in neural information processing systems, pp 802–810
55. Chatfield C (2004) The analysis of time series: an introduction. Chapman & Hall/CRC, Atlanta
56. Ertel W (2017) Introduction to artificial intelligence. Springer International Publishing, New York
57. Norvig P, Russell S (2016) Artificial intelligence: a modern approach, 3rd edn. Prentice Hall, Upper Saddle River
58. Yan WQ (2017) Introduction to intelligent surveillance: surveillance data capture, transmission, and analytics. Springer, Berlin
59. Chen J, Kang X, Liu Y, Wang Z (2015) Median filtering forensics based on convolutional neural networks. IEEE Signal Process Letters 22(11):1849–1853
60. Muscat J (2014) Functional analysis. Springer, Berlin
61. Hu X (2017) Frequency based texture feature descriptors. PhD thesis, Auckland University of Technology, New Zealand

Autoencoder and GAN

4

4.1 Autoencoder

Basic autoencoder [1–3] is a feedforward and non-recurrent neural network which is unsupervised learning. That means, our computers can learn from themselves.

Given a group of training data, how to encode the data and remove noises among them, these are typical applications of an autoencoder. Our objective of deep autoencoder is to reduce the dimension [4] and minimize the differences between the encoded data and the decoded data. Thus, autoencoder is a generative network, one of its advantages is to test the output as input and reduce the dimension of raw data [4]. Mathematically, for $\mathbf{x} \in \mathscr{R}^d$, $\mathbf{z} \in \mathscr{R}^p$,

$$\mathbf{z} = \sigma(\mathbf{W} \cdot \mathbf{x} + \mathbf{b}) \tag{4.1}$$

and

$$\mathbf{x}' = \sigma'(\mathbf{W}' \cdot \mathbf{z} + \mathbf{b}'). \tag{4.2}$$

To minimise the reconstruction errors, we have

$$L(\mathbf{x}, \mathbf{x}') = \|\mathbf{x} - \mathbf{x}'\|^2 = \|\mathbf{x} - \sigma'(W' \cdot \sigma(W \cdot \mathbf{x} + \mathbf{b}) + \mathbf{b}')\|^2. \tag{4.3}$$

Thus, the global loss function [5] is

$$J_{AE}(\Theta) = \sum_{\mathbf{x}} L(\mathbf{x}, \mathbf{x}'), \tag{4.4}$$

where $\Theta = (\mathbf{W}, \mathbf{b}, \mathbf{W}', \mathbf{b}')^\tau$. The decay equation is

$$\Theta_{i+1} := \Theta_i - \alpha \cdot \frac{\partial L_{AE}(\Theta_i)}{\partial \Theta_i}, \tag{4.5}$$

where $\alpha \geq 0$ is the learning rate.

© The Author(s), under exclusive license to Springer Nature Switzerland AG 2021
W. Q. Yan, *Computational Methods for Deep Learning*, Texts in Computer Science,
https://doi.org/10.1007/978-3-030-61081-4_4

4.2 Regularizations and Autoencoders

For L_2 regularization [6], if λ is the parameter of weight decay (wd), then

$$J_{AE+wd}(\Theta) = J_{AE}(\Theta) + \lambda \cdot \sum_{w_{i,j} \in \mathbf{W}} w_{i,j}^2. \tag{4.6}$$

For sparse regularization (sp) [7] using KL divergence [8], if β is the parameter of sparse weight, then

$$J_{AE+sp}(\Theta) = J_{AE}(\Theta) + \beta \cdot \sum_{j=1}^{m} KL(\rho||\hat{\rho}_j), \tag{4.7}$$

where

$$KL(\rho||\hat{\rho}_j) \overset{\Delta}{=} \rho \cdot \log \frac{\rho}{\hat{\rho}_j} + (1 - \rho) \cdot \log \frac{1 - \rho}{1 - \hat{\rho}_j} \tag{4.8}$$

and

$$\hat{\rho}_j = \frac{1}{N} \sum_{i=1}^{N} h_j(x^{(i)}) \tag{4.9}$$

$\hat{\rho}_j = \rho_j, j = 1, 2, \ldots, m; \mathbf{x} = \{x^{(i)}\}_{i=1}^{N}$. From the definition, we know that $KL(\rho||\hat{\rho}_j) \neq KL(\hat{\rho}_j||\rho)$. Furthermore,

$$J_{AE+wd+sp}(\Theta) = J_{AE}(\Theta) + \lambda \cdot \sum_{w_{ij} \in \mathbf{W}} w_{ij}^2 + \beta \cdot \sum_{j=1}^{m} KL(\rho||\hat{\rho}_j). \tag{4.10}$$

Thus,

$$J_{AE+wd+sp}(\Theta) = J_{AE+wd}(\Theta) + \beta \cdot \sum_{j=1}^{m} KL(\rho||\hat{\rho}_j) \tag{4.11}$$

and

$$J_{AE+wd+sp}(\Theta) = J_{AE+sp}(\Theta) + \lambda \cdot \sum_{w_{ij} \in \mathbf{W}} w_{ij}^2. \tag{4.12}$$

In denoising by using autoencoder [9], data corruption means

$$\mathbf{x}' = \mathbf{x} + \varepsilon, \tag{4.13}$$

where $\varepsilon \sim \mathbf{N}(\mu, \delta) \to \mathbf{N}(0, \delta^2\mathbf{I})$, $\mathbf{N}(0, \delta^2\mathbf{I})$ is the additive isotropic Gaussian noise. In contractive autoencoder (CAE),

$$J_{CAE}(\Theta) = J_{AE}(\Theta) + \lambda \cdot \|J_f\|^2, \tag{4.14}$$

where $\|J_f\|_F$ is Frobenius norm,

$$J_f = (a_{ij})_{m \times n} \overset{\Delta}{=} \left(\frac{\partial h_i}{x_j}\right)_{m \times n} \tag{4.15}$$

and

$$\|J_f\|_F^2 = \sum_{i=1}^{m} \sum_{j=1}^{n} \left(\frac{\partial h_i}{x_j} \right)^2, \tag{4.16}$$

where

$$h_i = \sigma (\mathbf{W} \cdot \mathbf{x} + \mathbf{b}). \tag{4.17}$$

Hence,

$$\|J_f\|_F^2 = \sum_{i=1}^{m} h_i \cdot (1 - h_i) \cdot \sum_{j=1}^{n} w_{ij}^2. \tag{4.18}$$

For variational autoencoder (VAE) [10, 11], we define Kullback–Leibler (KL) divergence [8] as

$$KL(Q||P) \triangleq \sum_{x \in \mathbf{x}} Q(x) \cdot \log \frac{Q(x)}{P(x)} \triangleq \int_{x \in \mathbf{x}} Q(x) \frac{Q(x)}{P(x)} dx. \tag{4.19}$$

Bayes' Theorem [12] is

$$P(x|z) = \frac{P(z|x) \cdot P(x)}{P(z)} = \frac{P(z|x) \cdot P(x)}{\sum_{x \in \mathbf{x}} P(z|x) \cdot P(x)}, \tag{4.20}$$

where the probabilities $P(x|z)$, $P(z|x)$, $P(x)$, and $P(z)$ refer to posterior, likelihood, prior, and evidence, respectively. Therefore, we have variational inference [12]. If $Q(z) = P(x|z)$, then

$$KL(Q(z)||P(z|x)) = KL(Q(z)||P(z)) - \sum_{z \in \mathbf{z}} Q(z) \log P(x|z) + \log P(x). \tag{4.21}$$

Now, we define VAE. If $x \sim \mathbf{N}(\mu, \delta)$, $z = x + \varepsilon = g(x, \varepsilon)$, then $z \sim \mathbf{N}(\mu + \varepsilon, \delta)$, $\mathbf{N}(\mu, \delta)$ is the normal or Gaussian distribution, μ is mean, δ is variance,

$$Q(z) = P(x|z) = P(\varepsilon). \tag{4.22}$$

Furthermore,

$$L_{VAE}(Q, P) = KL(Q(z)||P(z|x)) \tag{4.23}$$

$$= KL(Q(z)||P(z)) - \sum_{\mathbf{z} \sim N(\mu + \varepsilon, \delta)} Q(z) \log P(x|z) + \log P(x) \tag{4.24}$$

$$= KL(Q(z)||P(z)) - \sum P(\varepsilon) \log P(x|g(x, \varepsilon)) + \log P(x), \tag{4.25}$$

where $g(x, \varepsilon)$ is an encoder model, $P(x|z)$ is a decoder model. Hence, the cost function is

$\min[KL(Q(z)||P(z|x))] \Leftrightarrow$

$$\underbrace{\min KL(Q(z)||P(z))}_{Encoder:KL\,Divergence} - \underbrace{\max \sum_{\mathbf{z} \sim N(\mu + \varepsilon, \delta)} Q(z) \log P(x|z)}_{Decoder:Max\,Likelihood}, \tag{4.26}$$

where $\log P(x)$ w.r.t. $x \in \mathbf{x}$ is independent on $z \in \mathbf{z}$ [12].

In a nutshell, a VAE consists of an encoder, a decoder, and a loss function. The term "variational" comes from the close relationship between the regularisation and the variational inference method in statistics. VAE outputs a Gaussian probability distribution with mean and standard deviation for every dimension. For a given set of possible encoders and decoders, VAEs look for the pair that keeps the maximum of information when encoding and has the minimum of reconstruction error when decoding. VAE is trained by using gradient descent to optimize the loss with respect to the parameters of the encoder and decoder.

Autoencoders are a type of self-supervised learning model that can learn a representation through input data. An LSTM Autoencoder [13] is an implementation of an autoencoder for sequence data using an Encoder–Decoder LSTM architecture. LSTM Autoencoders can learn a representation of sequence data. For a given dataset of sequences, an encoder–decoder LSTM is configured to read the input sequence, encode it, decode it, and recreate it. The performance of LSTM autoencoder is evaluated based on the model's ability to recreate the input sequence.

Autoencoders have been applied to remove image noises, such as haze removal; it eventually is employed to implement image inpainting such as TV logo removal because the output could be used as the input, iteratively.

In MATLAB, an autoencoder is a neural network which replicates its input as its output. When the number of neurons in the hidden layer is less than the size of the input, autoencoder learns a compressed representation of the input. This autoencoder uses regularizers to learn a sparse representation in the first layer. The regularizers are controlled by setting various parameters. A MATLAB example of training autoencoders for image classification is available.

4.3 Generative Adversarial Networks

Generative adversarial network (GAN) learns to generate new data with the same statistics as the given training set. The generative network generates candidates, while the discriminative network evaluates them. The generator is typically a deconvolutional neural network, the discriminator is a convolutional neural network [14, 15]. GAN is applied to digital forensics and find the real one from fake ones.

Given $\forall\{x_1, x_2, \ldots, x_m\} \sim P_{data}(x)$, $P_{data}(x) \approx P_{G(x,\Theta)}$, the maximum likelihood estimation of x_i in $P_{G(x,\Theta)}$ is

$$L = \prod_{i=1}^{m} P_G(x_i, \Theta). \tag{4.27}$$

For the parametric optimization,

$$\Theta^* = \arg \max_{\Theta} \prod_{i=1}^{m} \mathbf{P}_G(x_i, \Theta). \tag{4.28}$$

Hence,

$$\Theta^* = \arg\max_{\Theta} \sum_{i=1}^{m} \log \mathbf{P}_G(x_i, \Theta). \tag{4.29}$$

Furthermore,

$$\Theta^* = \arg\max_{\Theta} KL(P_{data}(x)||P_G(x, \Theta)) \tag{4.30}$$

- **Generator** G: Generate x from z.
- **Discriminator** D: Evaluate the difference between $P_{data}(x)$ and $P_G(x, \Theta)$ through function

$$V(G, D) \overset{\Delta}{=} \mathbf{E}_{x \sim P_{data}}(\log D(x)) + \mathbf{E}_{x \sim P_G}(\log(1 - D(x))) \tag{4.31}$$

$$G^* = \arg\min_{G}\max_{D} V(G, D). \tag{4.32}$$

Given G, if

$$D^*(x) = \frac{P_{data}(x)}{P_{data}(x) + P_G(x)}, \tag{4.33}$$

because

$$V(G, D^*) = \max V(G, D) = -2\log 2 + 2JS(P_{data}(x)||P_G(x)). \tag{4.34}$$

Jensen–Shannon (JS) divergence is

$$JS(P||Q) = \frac{1}{2}KL(P||M) + \frac{1}{2}KL(Q||M), \tag{4.35}$$

where $M = \frac{P+Q}{2}$ and $JS(P||Q) = JS(Q||P)$, however, $KL(P||Q) \neq KL(Q||P)$.
Thus,

$$G^* = \arg\min_{G}\max_{D^*} V(G, D^*) \tag{4.36}$$

$$L(G) = \max_{D^*} V(G, D^*), \tag{4.37}$$

therefore,

$$G^* = \arg\min_{G} L(G). \tag{4.38}$$

Hence,

$$\Theta_G := \Theta_G - \beta \cdot \frac{\partial L(G)}{\partial \Theta_G}, \tag{4.39}$$

where $\beta \geq 0$ is the learning rate. We solve this problem by using the following way:

- Given G_0,

$$D_1 = \arg\max_{D} V(G_0, D) \tag{4.40}$$

- Given D_1,

$$G_1 = \arg\max_{G} V(G, D_1). \tag{4.41}$$

- $\bullet$ $\cdots\cdots$
- $G_i \Rightarrow D_{i+1}; D_{i+1} \Rightarrow G_{i+1}.$
- $\bullet$ $\cdots\cdots$

$$G^* = \arg \min_G \max_D V(G, D), \tag{4.42}$$

where

$$V = \mathbf{E}_{x \sim P_{data}}[\log D(x)] + \mathbf{E}_{x \sim P_G}[\log(1 - D(x))]. \tag{4.43}$$

Discretely,

$$V = \frac{1}{m} \cdot \sum_{i=1}^{m} \log D(x_i) + \frac{1}{m} \cdot \sum_{i=1}^{m} \log[1 - D(\hat{x}_i)] \tag{4.44}$$

where $x_i \sim P_{data}$ and $\hat{x}_i \sim P_G$. We thus have

$$\Theta_D := \Theta_D - \beta \cdot \frac{\partial V}{\partial \Theta_D}. \tag{4.45}$$

If $z_i \sim N(0, 1)$, $\hat{x}_i = G(z_i)$, then,

$$V = \frac{1}{m} \cdot \sum_{i=1}^{m} \log[1 - D(G(z_i))]. \tag{4.46}$$

Thus, we have

$$\Theta_G := \Theta_G - \beta \cdot \frac{\partial V}{\partial \Theta_G}. \tag{4.47}$$

Therefore, we use Eqs. (4.46) and (4.47) to implement GAN. If we set image processing as an example, GAN can make a picture clear using existing details like superresolution, GAN also can remove artefacts of digital images, etc. MATLAB has an example of how to train GAN models available.

SimGAN [15] refines the output of the simulator of a neural network. We need to minimize the image difference between the synthetic one and the refined images, and finally, update the discriminator alternately.

SimGan uses unlabelled real data to refine the synthetic images, trains a refined network to add random numbers to synthetic images, further stabilizes GAN training, and prevents the refiner network from producing artefacts as well as generates the results without human annotation by training deep neural networks on the refined output images. The overall loss function is

$$L_R(\theta) = \sum_i X_i l_{real}(\theta; \mathbf{x}_i, L) + \lambda l_{reg}(\theta; \mathbf{x}_i), \tag{4.48}$$

where

$$l_{real}(\theta; \mathbf{x}_i, L) = -\log(1 - D_\phi(R_\theta(\mathbf{x}_i))) \tag{4.49}$$

and

$$l_{reg}(\theta; \mathbf{x}_i) = \|\psi(\tilde{\mathbf{x}}) - \mathbf{x}\|, \tag{4.50}$$

where $y_i \in \mathbf{y}$ is an unlabelled real image, $x_i \in \mathbf{x}$ is a synthetic training image.

The discriminator updates its parameters by minimizing the loss function

$$L_D(\phi) = -\sum_i \log(D_\phi(x_i)) - \sum_j \log(1 - D_\phi(x_i)), \qquad (4.51)$$

where x_i is a synthetic image.

4.4   Information Theory

Text processing is employed in natural language processing. Text information has entropy, the information capacity is measured by using entropy, even the text short message (SMS) only has 144 letters that could be measured by entropy.

$$H = -\sum_{i=1}^{m} p_i \ln p_i = \mathbf{E} \ln \frac{1}{p_i} \qquad (4.52)$$

where H is entropy, $p_i \in [0, 1]$ is the probability, it may be the histogram of 256 letters (ASCII code) or pixels with 256 grayscale intensities after the normalization.

Probability is usually between 0 and 1, v.i.z., $p \in [0, 1]$. Entropy could be written in the way of mathematical expectation $\mathbf{E}(\cdot)$; correspondingly, we define joint entropy, conditional entropy, relevant entropy.

We denote conditional probability as $p(x|y)$. Given x, the entropy $h(x|y)$ is not as same as the entropy given y. Therefore, we define the joint entropy and mutual entropy.

Joint probability is $h(x, y)$. Information capacity is defined as $c = \max I(x, y)$. In the Internet, information theory and entropy have their broad applications.

Relevant entropy is also called KL divergence between p and q, which reflects the information distance between p and q, $KL(p||q) = -\sum_{i=1}^{m} p_i \ln \frac{p_i}{q_i}$. Again, $KL(p||q) \neq KL(q||p)$.

Mutual information is defined based on joint probability. KL divergence has been used in deep learning for entropy-based loss functions and distance computing.

In graphical models, we use relevant entropy, joint entropy, and mutual information. The mutual information has multiple definitions, they equal to each other, and could be written in the product form if the probability of each element is independent.

Joint entropy and mutual entropy are shown in Venn diagram (also called primary diagram, set diagram or logic diagram), which shows all possible logical relations between a finite collection of different sets. Regarding these concepts of entropy, we have the chain rule for the case of infinity. Correspondingly, we use conditional entropy.

According to Bayes' theorem, we have the relationship between joint entropy, conditional entropy, relevant entropy, and mutual information.

Jassen's inequality tells us what a convex or concave function is. If the points on a curve are always located at one side of the straight line, we say the curve is convex. Mathematically,

$$f(\alpha \cdot x_1 + (1 - \alpha)x_2) \leq \alpha \cdot f(x_1) + (1 - \alpha) \cdot f(x_2), \alpha \in [0, 1]. \qquad (4.53)$$

If a function is convex, the second derivative could be applied to decide whether it is such a function. $f'(x) \geq 0$ and $f''(x) \geq 0$.

Suppose we have $p(x)$ and $q(x)$, for any two functions, we have the relevant entropy

$$H(p||q) = -p \ln \frac{p}{q}. \tag{4.54}$$

Another concept is entropy rate

$$H(p) = -\frac{1}{n} \sum_{i=1}^{m} \ln p_i. \tag{4.55}$$

Generally, the entropy $H(X)$ of a discrete random variable X is defined by using

$$H(X) = -\sum_{x \in X} p(x) \log p(x) = -\mathbf{E} \log p(X) = \mathbf{E} \log \frac{1}{p(X)}. \tag{4.56}$$

The joint entropy $H(X, Y)$ of a pair of discrete random variables (X, Y) with a joint distribution $p(x, y)$ is defined as

$$H(X, Y) = -\sum_{x \in X, y \in Y} p(x, y) \log p(x, y) = \mathbf{E} \log \frac{1}{p(X, Y)}. \tag{4.57}$$

If $(X, Y) \sim p(x, y)$, then conditional entropy [8] $H(Y|X)$ is

$$H(Y|X) = \sum_{x \in X} p(x) H(Y|X = x) \tag{4.58}$$

$$= -\sum_{x \in X} p(x) \sum_{y \in Y} p(y|x) \log p(y|x) = \mathbf{E}_{p(x,y)} \log \frac{1}{p(Y|X)}. \tag{4.59}$$

Corollary,

$$H(X|Y) \neq H(Y|X). \tag{4.60}$$

Equivalently, we denote

$$\log p(X, Y) = \log p(X) + \log p(Y|X). \tag{4.61}$$

For the chain rule of entropy,

$$H(X, Y) = H(X) + H(Y|X) \tag{4.62}$$

and

$$H(X, Y|Z) = H(X|Z) + H(Y|X, Z). \tag{4.63}$$

Mutual information [8] $I(X; Y)$ is a measure of the dependence between two random variables, which is symmetric and always nonnegative,

$$I(X; Y) = H(X) - H(X|Y) \tag{4.64}$$

and

$$I(X; Y) = H(X) - H(X|Y) = H(Y) - H(Y|X). \tag{4.65}$$

For a communication channel with input X and output Y, the capacity C is defined,

$$C = \max_{p(x)} I(X; Y) \tag{4.66}$$

The capacity is the maximum rate at which we send information over the channel and recover the information at the output with a vanishingly low probability of error.

Relative entropy (KL Divergence) [8] is a measure of the "distance" between two probability mass functions p and q.

$$D(p\|q) = \sum_{x \in X} p(x) \log \frac{p(x)}{q(x)} = \mathbf{E}_{p(X)} \log \frac{p(X)}{q(X)}, \tag{4.67}$$

where $D(p\|q) \neq D(q\|p)$. Mutual information [8] $I(X; Y)$ is the relative entropy between the joint distribution and the product distribution $p(x)q(x)$.

$$I(X; Y) = \sum_{x \in X} \sum_{y \in Y} p(x, y) \log \frac{p(x, y)}{p(x)p(y)} \tag{4.68}$$

$$= D(p(x, y)\|p(x)p(y)) = \mathbf{E}_{p(x,y)} \log \frac{p(X, Y)}{p(X)p(Y)}. \tag{4.69}$$

Meanwhile,

$$I(X; Y) = H(X) - H(X|Y) = H(Y) - H(Y|X) \tag{4.70}$$

$$I(X; Y) = H(X) + H(Y) - H(X, Y). \tag{4.71}$$

According to Bayes' theorem [12],

$$p(x_1, x_2) = p(x_2)p(x_1|x_2) = p(x_1)p(x_2|x_1) \tag{4.72}$$

$$H(X_1, X_2) = H(X_1) + H(X_2|X_1) \tag{4.73}$$

$$H(X_1, X_2, X_3) = H(X_1) + H(X_2, X_3|X_1) = H(X_1) + H(X_2|X_1) + H(X_3|X_2, X_1). \tag{4.74}$$

......

Therefore,

$$H(X_1, X_2, X_3, \ldots, X_n) = \sum_{i=1}^{N} H(X_i|X_{i-1}, \ldots, X_2, X_1). \tag{4.75}$$

Chain rule for relative entropy,

$$D(p(x, y)\|q(x, y)) = D(p(x)\|q(x)) + D(p(y|x)\|q(y|x)), \tag{4.76}$$

where

$$D(p(y|x)\|q(y|x)) = \sum_{x} p(x) \sum_{y} p(y|x) \log \frac{p(y|x)}{q(y|x)} = \mathbf{E} \log \frac{p(y|x)}{q(y|x)}. \tag{4.77}$$

If $f(\cdot)$ is a convex function and X is a random variable, then

$$\mathbf{E} f(X) \geq f(\mathbf{E}(X)). \tag{4.78}$$

A function $f(x)$ is convex over an interval (a, b) if for every $x_1, x_2 \in (a, b)$ and $0 \leq \lambda \leq 1$,

$$f[\lambda x_1 + (1 - \lambda)x_2] \leq \lambda f(x_1) + (1 - \lambda)f(x_2). \tag{4.79}$$

If $f(\cdot)$ is a convex function and X is a random variable, then

$$\mathbf{E} f(X) \geq f(\mathbf{E}(x)). \tag{4.80}$$

If the function $f(\cdot)$ has a second derivative $f''(x) \geq 0$ everywhere, then the function is convex (strictly convex). Let $p(x), q(x), x \in \mathbf{X}$ be two probability mass functions, then $D(p\|q) \geq 0$, $D(p(x|y)\|q(x|y)) \geq 0$.

For any two random variables X and Y, $I(X; Y) \geq 0$ and $I(X; Y|Z) \geq 0$; furthermore, because $I(X; Y) = H(X) - H(X|Y) \geq 0$, $H(X) \geq H(X|Y)$.

$$H(X_1, X_2, \ldots, X_n) = \sum_{i=1}^{N} H(X_i|X_{i-1}, \ldots, X_1) \leq \sum_{i=1}^{N} H(X_i). \tag{4.81}$$

For $a_i, b_i \geq 0, i = 1, 2, \ldots, n$

$$\sum_{i=1}^{N} a_i \log \frac{a_i}{b_i} \geq \sum_{i=1}^{N} a_i \log \frac{\sum_{i=1}^{N} a_i}{\sum_{i=1}^{N} b_i} \tag{4.82}$$

$$D(p\|q) = \sum_{i=1}^{N} p(x) \log \frac{p(x)}{q(x)} \geq \sum p(x) \log \frac{\sum p(x)}{\sum q(x)}. \tag{4.83}$$

Hence,

$$\lambda D(p_1\|q_1) + (1 - \lambda)D(p_2\|q_2) \geq D(\lambda p_1 + (1 - \lambda)p_2\|\lambda q_1 + (1 - \lambda)q_2). \tag{4.84}$$

Therefore, $H(X)$ is a convex function and $I(X; Y)$ is a concave function [16]. The entropy rate of a stochastic process X_i is defined by

$$H(\mathbf{X}) = \lim_{n \to \infty} \frac{1}{n} H(X_1, X_2, \ldots, X_n). \tag{4.85}$$

A related quantity for entropy rate [8]:

$$H'(\mathbf{X}) = \lim_{n \to \infty} \frac{1}{n} H(X_n|X_{n-1}, \ldots, X_1). \tag{4.86}$$

For a stationary stochastic process,

$$H'(\mathbf{X}) = H(\mathbf{X}) \Rightarrow \tag{4.87}$$

$$\lim_{n \to \infty} H(X_n|X_{n-1}, \ldots, X_1) = \lim_{n \to \infty} H(X_n|X_{n-1}) = H(X_2|X_1). \tag{4.88}$$

Let $X_i, i = 1, 2, \ldots$ be a stationary Markov chain with stationary distribution μ and transition matrix $\mathbf{P} = (P_{ij})$. Then the entropy rate is

$$H(\mathbf{X}) = -\sum_{ij} \mu_i P_{ij} \log P_{ij}. \tag{4.89}$$

The entropy rate of the two states Markov chain is

$$H(\mathbf{X}) = H(X_2|X_1) = \frac{\alpha}{\alpha + \beta} \cdot H(\beta) + \frac{\beta}{\alpha + \beta} \cdot H(\alpha). \tag{4.90}$$

Entropy is defined not only in a discrete way but also in a continuous way. Previously, it was based on the sum function, now it is based on the integral operation. Previously, the entropy was H, now it is h. If $f(\cdot)$ is continuous function, the entropy function will be continuity. The continuous entropy is

$$h = -\int p(x) \ln p(x) dx = -\mathbf{E} \ln \frac{1}{p(x)}. \tag{4.91}$$

The continuous conditional entropy is

$$h = -\int p(x|y) \ln p(x|y) dx = -\mathbf{E} \ln \frac{1}{p(x|y)}. \tag{4.92}$$

The continuous joint entropy is

$$h = -\int p(x, y) \ln p(x, y) dx = -\mathbf{E} \ln \frac{1}{p(x, y)}. \tag{4.93}$$

The continuous entropy rate is

$$h = -\frac{1}{L} \int p(x, y) \ln p(x, y) dx = -\frac{1}{L} \mathbf{E} \ln \frac{1}{p(x, y)}. \tag{4.94}$$

4.5 Questions

Question 1. How does the autoencoder generate images which look similar to the original image?

Question 2. What are the relationships between autoencoder and GAN?

Question 3. How to measure the performance of the generator and discriminator in GAN?

Question 4. What's the chain rule for relative entropy?

Question 5. What are the advantages and disadvantages using Relative entropy (KL) Divergence as a measure between two probability mass functions? How to solve this problem?

References

1. Xing C, Ma L, Yang X (2016) Stacked denoise autoencoder based feature extraction and classification for hyperspectral images. J Sens 2016
2. Masci J, Meier U, Cirean D, Schmidhuber J (2011) Stacked convolutional autoencoders for hierarchical feature extraction. In: International conference on artificial neural networks. Springer, Berlin, pp 52–59
3. Wang J, Zhang C (2018) Software reliability prediction using a deep learning model based on the RNN encoder - decoder. Reliab Eng Syst Saf 170:73–82
4. Hinton GE, Salakhutdinov RR (2006) Reducing the dimensionality of data with neural networks. Science 313(5786):504–507
5. Ko YH, Kim KJ, Jun CH (2005) A new loss function-based method for multiresponse optimization. J Qual Technol 37(1):50–59
6. Wan L, Zeiler M, Zhang S, Le Cun Y, Fergus R (2013) Regularization of neural networks using DropConnect. In: International conference on machine learning, pp 1058–1066
7. Poultney C, Chopra S, Cun YL (2007) Efficient learning of sparse representations with an energy-based model. In: Advances in neural information processing systems, pp 1137–1144
8. Cover T, Thomas J (1991) Elements of information theory. Wiley, New York
9. Li CP, Qin PY, Zhang JJ (2017) Research on image denoising based on deep convolutional neural network. Comput Eng 43(3)
10. Marreiros AC, Daunizeau J, Kiebel SJ, Friston KJ (2008) Population dynamics: variance and the sigmoid activation function. Neuroimage 42(1):147–157
11. Welling M, Kingma D (2019) An introduction to variational autoencoders. Found Trends Mach Learn 12(4):307–392
12. Koller D, Friedman N (2009) Probabilistic graphical models. MIT Press, Cambridge
13. Marchi E, Vesperini F, Squartini S, Schuller B (2017) Deep recurrent neural network-based autoencoders for acoustic novelty detection. Comput Intell Neurosci Hindawi (Article ID 4694860)
14. Ng AY, Jordan MI (2002) On discriminative vs. generative classifiers: a comparison of logistic regression and Naive Bayes. In: Advances in neural information processing systems, pp 841–848
15. Shrivastava A et al. Learning from simulated and unsupervised images through adversarial training, In: CVPR'17
16. Muscat J (2014) Functional analysis. Springer, Berlin

Reinforcement Learning

5

5.1 Introduction

Learning from the interaction is a foundational idea underlying nearly all theories of learning and intelligence. Reinforcement learning is learning about how to map situations to actions. Actions may affect not only the immediate reward, but also the next situation and all subsequent rewards. Reinforcement learning is related to dynamical systems, specifically, optimal control and Markov decision process (MDP). Reinforcement learning explicitly considers the whole problem of a goal-directed agent interacting with an uncertain environment. Reinforcement learning seeks the tradeoff between exploration and exploitation.

A reinforcement learning system includes a policy, a reward signal, a value function, and a model of the environment. A policy defines the learning agent's way of behaving at a given time. A reward signal defines the goal in a reinforcement learning problem. A value function specifies what is good in the long run. A value of a state is the total amount of reward an agent can expect to accumulate over the future, starting from that state. Rewards determine the immediate, intrinsic desirability of environmental states. The final element of reinforcement learning is a model of the environment.

Google street view could navigate us in an outdoor environment, but within a building, how could Google street view assist us? Reinforcement learning [1, 2] could navigate us in this indoor environment. Reinforcement learning is one of three basic machine learning paradigms, alongside supervised learning and unsupervised learning. Assume we have a building map, how could a robot lead us to leave this building or find a room in this building? Successfully solving this problem can assist us to find the shortest path in a shopping mall or underground without GPS information quickly. In a university, it also could quickly help students to find their meeting rooms or classrooms and rapidly aid a robot to get the destination in an indoor environment.

In summary, reinforcement learning relies heavily on the state and the policy, which is a computational approach to understand and automate goal-directed learning

© The Author(s), under exclusive license to Springer Nature Switzerland AG 2021
W. Q. Yan, *Computational Methods for Deep Learning*, Texts in Computer Science,
https://doi.org/10.1007/978-3-030-61081-4_5

and decision-making. Reinforcement learning uses Markov decision processes to define the interaction between a learning agent and its environment in terms of states, actions, and rewards.

5.2 Bellman Equation

The research problems in reinforcement learning include the bandit problems, finite Markov decision problem, Bellman equations and value functions. The Bellman optimality equations are special consistency conditions, from which an optimal policy can be determined with relative ease. In finite markov decision problems, the research methods such as dynamic programming, Monte Carlo methods, and temporal-difference learning methods are taken into consideration.

We call a software robot as the agent, which has the intelligent capability to make decision by itself, the agent is living in an environment. We need an environment with a policy, actions, related rewards, effects, punishment, and states. States are thought as the input of a policy and actions. The best policy and the best reward [3] are obtained by using optimization.

Assume we have an agent and an environment, we denote the action of an agent a, a reward r, a policy π, a state s and an action, which is defined by policy and state, i.e., $a \stackrel{\Delta}{=} \pi(s)$. We denote the samples as $\{s_1, a_1, r_1, \ldots, s_t\}, t = 1, 2, \ldots$; therefore, reinforcement learning is to find $\max(r)$, s.t. $\{s_1, a_1, r_1, \ldots, s_t\} \to \pi$.

A finite MDP is an MDP with finite state, action, and reward sets. The return is the function of future rewards that the agent seeks to maximize in expected value.

Markov decision process (MDP) only affects the next time, which does not affect too much of the sequence at present [4], that means,

- A state s_t is Markov if $P(s_{t+1}|s_t) = P(s_{t+1}|s_1, \ldots, s_t)$.
- Value function $v(s) \stackrel{\Delta}{=} \mathbf{E}(G_t|s_t)$
- Return $G_t \stackrel{\Delta}{=} \sum_{k=0}^{\infty} \lambda^k \cdot r_{t+k+1}$, λ is the discount factor.

Action-value function is defined as

$$Q^\pi(s, a) \stackrel{\Delta}{=} \mathbf{E}_{s'}(r + \lambda \cdot Q^\pi(s', a')|s, a). \tag{5.1}$$

The optimal action-value function therefore is

$$Q^*(s, a) = \mathbf{E}_{s'}(r + \lambda \cdot \max_{a'} Q^*(s', a')|s, a). \tag{5.2}$$

Iteratively,

$$Q_{i+1}(s, a) = \mathbf{E}_{s'}(r + \lambda \cdot \max_{a'} Q_i(s', a')|s, a) \to Q^*(i \to \infty), \tag{5.3}$$

where $\mathbf{E}(\cdot)$ is probability expectation, we call Eq. (5.3) as Bellman equation.

Rewards are decided by actions. Action-value function $Q(a, s)$ is defined by actions and states. The best Q is dependent on both action a and state s. The process to find the best Q is called the Q-learning. We use Q-learning to find the optimized policy and maximize the reward. Q-learning is a simplified Bellman equation, if $\alpha \in [0, 1]$,

$$Q(s_t, a_t) \leftarrow Q(s_t, a_t) + \alpha(r_{t+1} + \lambda \cdot \max_a Q(s_{t+1}, a) - Q(s_t, a_t)). \quad (5.4)$$

The best policy, state, and reward are associated with each other. For a deep network w.r.t. Q and weight w,

$$Q(s, a, w) = Q^\pi(s, a). \quad (5.5)$$

Thus, the loss or objective function is

$$L(w) = \mathbf{E}([r + \gamma \cdot \max_{a'} Q(s', a', w) - Q(s, a, w)]^2). \quad (5.6)$$

The gradient is

$$\frac{\partial L(w)}{\partial w} = \mathbf{E}([r + \gamma \cdot \max_{a'} Q(s', a', w) - Q(s, a, w)]) \cdot \frac{\partial Q(s, a, w)}{\partial w}. \quad (5.7)$$

Reinforcement learning is the use of value functions to organize and structure the search for good policies. Monte Carlo methods (MC) are ways of solving the reinforcement learning problem based on averaging sample returns. Each occurrence of state in an episode is called a visit. The first time visited in an episode is called as the first visit. The first-visit MC method estimates the average of the returns following first visits.

The policy iteration of MC methods is natural to alternate between evaluation and improvement on an episode-by-episode basis. After each episode, the observed returns are used for policy evaluation, and then the policy is improved at all the states visited in the episode. In MC methods, we use first-visit MC methods to estimate the action-value function for the current policy.

With MC methods, one must wait until the end of an episode, because only then the return is known, whereas with temporal-difference (TD) learning methods, one need wait only one time step. TD methods can learn directly from raw experience without a model of the environmental dynamics, which update estimates based on other learned estimates, without waiting for a final outcome.

The simplest TD method makes the update immediately on transition to S_{t+1} and receiving R_{t+1}.

$$V(S_t) \leftarrow V(S_t) + \alpha[R_{t+1} + \gamma \cdot V(S_{t+1}) - V(S_t)]. \quad (5.8)$$

TD error measures the difference between the estimated value of S_t and the better estimate $\delta_t \doteq R_{t+1} + \gamma V(S_{t+1}) - V(S_t)$.

In gradient descent methods, $\mathbf{w} = (w_1, \ldots, w_d)^\tau$,

$$\mathbf{w}_{t+1} = \mathbf{w}_t + \alpha[v_\pi(S_t) - \hat{v}(S_t, \mathbf{w}_t)]\nabla\hat{v}(S_t, \mathbf{w}_t), \quad (5.9)$$

where α is a positive step-size parameter, $\nabla\hat{v}(S_t, \mathbf{w}_t)$ is the gradient with respect to $\mathbf{w}$. This yields the following general SGD method for state-value prediction:

$$\mathbf{w}_{t+1} = \mathbf{w}_t + \alpha[U_t - \hat{v}(S_t, \mathbf{w}_t)]\nabla\hat{v}(S_t, \mathbf{w}_t). \quad (5.10)$$

Because the true value of a state is the expected value of the return, the MC target $U_t \doteq G_t$.

Linear methods approximate the state-value function by using inner product

$$\hat{v}(s, \mathbf{w}) = \mathbf{w}^\tau \mathbf{x}(s) = \sum_{i=1}^{d} w_i x_i(s),$$

where $\hat{v}(\cdot, \mathbf{w})$ is a linear function of the weight vector $\mathbf{w}$, $\mathbf{x}(s) = (x_1(s), x_2(s), \ldots, x_d(s))^\tau$ is a real-valued vector, $\mathbf{x}(s)$ is called a feature vector representing state s.

The gradient of the approximate value function is $\nabla \hat{v}(s, \mathbf{w}) = \mathbf{x}(s)$. The general SGD update is

$$\mathbf{w}_{t+1} = \mathbf{w}_t + \alpha [U_t - \hat{v}(S_t, \mathbf{w})] \mathbf{x}(s_t). \tag{5.11}$$

The gradient of the approximate value function is $\nabla \hat{v}(s, \mathbf{w}) = \mathbf{x}(s)$. The general SGD update is

$$\mathbf{w}_{t+1} = \mathbf{w}_t + \alpha [U_t - \hat{v}(S_t, \mathbf{w})] \mathbf{x}(s_t). \tag{5.12}$$

For example:

$$\mathbf{w}_{t+n} = \mathbf{w}_{t+n-1} + \alpha [G_{t:t+n} - \hat{v}(S_t, \mathbf{w}_{t+n-1})] \nabla \hat{v}(S_t, \mathbf{w}_{t+n-1}),$$

where $G_{t:t+n} = R_{t+1} + \gamma \cdot R_{t+2} + \cdots + \gamma^{n-1} \cdot R_{t+n} + \gamma^n \cdot \hat{v}(S_{t+n}, \mathbf{w}_{t+n-1})$, $0 \le t \le T - n$.

5.3 Deep Q-Learning

Reinforcement learning could learn the best policy and maximize the total reward. The key of reinforcement learning is to maximize the rewards, the question is how to get the best action so as to achieve the best reward. The sequence of actions thus will have the maximum cumulative reward. For each policy π, there is a reward $v^\pi(s_t)$, we hope to find the optimal policy

$$v^*(s_t) = \max_{\pi}(v^\pi(s_t)), \forall s_t. \tag{5.13}$$

In a simple case, action $a(t) \stackrel{\Delta}{=} \pi(s_t)$, $Q(a_t) = r(a_t) > 0$. If $r(a)$ is the reward

$$Q(a_{t+1}) \leftarrow Q_t(a_t) + \eta \cdot [r(a_{t+1}) - Q(a_t)]. \tag{5.14}$$

In a full reinforcement learning, a policy π defines the action to be taken in any state

$$a_t \stackrel{\Delta}{=} \pi(s_t). \tag{5.15}$$

The value of state s_t satisfies

$$v(s_t) = \max_{a_t} Q(s_t, a_t), \tag{5.16}$$

$$a_t^* = \arg\max_{a_t} Q(s_t, a_t), \tag{5.17}$$

and

$$\pi^*(s_t^*) = a_t^*. \tag{5.18}$$

The value iteration is

$$|v^{(l+1)}(s) - v^{(l)}(s)| < \delta, \tag{5.19}$$

where $\delta > 0; l = 1, 2, 3, \ldots$ and

$$v(s_t) \leftarrow v(s_t) + \eta \cdot [r_{t+1} + \gamma \cdot v(s_{t+1}) - v(s_t)]. \tag{5.20}$$

The policy iteration is

$$\pi \leftarrow \pi' = \arg\max_{\pi}(v^\pi(s')); \tag{5.21}$$

and

$$v^\pi(s) \leftarrow v^\pi(s'). \tag{5.22}$$

The rewards and actions are

$$Q(a_t, s_t) = r_{t+1} + \gamma \cdot \max_{a_{t+1}} Q(a_{t+1}, s_{t+1}). \tag{5.23}$$

This iteration has been employed to approximate the best value. Hence, we develop the iteration further:

- **Episode**: $\exists T, (s_1, a_1, r_2, \ldots, s_T) \to \pi$
- **Monte-Carlo Method**: using empirical mean to replace Bellman equation instead of expected return, i.e.,

$$v_\pi(s) = \frac{1}{T} \sum_{t=1}^{T} (G_t | s_t = s), \tag{5.24}$$

where $G_t = \sum_{k=1}^{T-t} \lambda^{k-1} r_{t+k}$. Hence,

$$\pi(s) \leftarrow \arg\max_{a} Q(s, a) \tag{5.25}$$

$$v(s_t) \leftarrow v(s_t) - \alpha \cdot (G_t - v(s_t)) \tag{5.26}$$

- Temporal Difference (TD):

$$v(s_t) \leftarrow v(s_t) - \alpha \cdot (r_{t+1} + \gamma \cdot v(s_{t+1}) - v(s_t)), \tag{5.27}$$

where $r_{t+1} + \gamma \cdot v(s_{t+1}) - v(s_t)$ is the TD error and $r_{t+1} + \gamma \cdot v(s_{t+1})$ is the TD target.

For the best convergence, we use Q-learning and double Q-learning for finding the best policies and actions:

- **Episode**: $\exists T, (s_1, a_1, r_2, \ldots, s_T) \rightarrow \pi$
- **SARSA (State-Action-Reward-State-Action) Algorithm**:

$$Q(s, a) \leftarrow Q(s, a) + \alpha \cdot [r + \gamma \cdot Q(s', a') - Q(s, a)], \qquad (5.28)$$

where $s \leftarrow s'$ and $a \leftarrow a'$.
- **Q-Learning:** An off-policy TD control algorithm.

$$Q(s, a) \leftarrow Q(s, a) + \alpha \cdot [r + \gamma \cdot \max_a Q(s', a) - Q(s, a)], \qquad (5.29)$$

where $s \leftarrow s'$
- **Double Q-Learning:**

$$Q_1(s, a) \leftarrow Q_1(s, a) + \alpha \cdot [r + \gamma \cdot Q_2(s', \arg \max_a Q_1(s', a)) - Q_1(s, a)] \quad (5.30)$$

and

$$Q_2(s, a) \leftarrow Q_2(s, a) + \alpha \cdot [r + \gamma \cdot Q_1(s', \arg \max_a Q_2(s', a)) - Q_2(s, a)], \quad (5.31)$$

where $s \leftarrow s'$.

The control is very similar to Kalman filtering, but Kalman filtering is a linear dynamic system for signal filtering.

Reinforcement learning enables a computer to make a series of decisions to maximize the cumulative reward for the task without human intervention and without being explicitly programmed to achieve the task. MATLAB lists all examples of reinforcement learning.

An example to swing up and balance pendulum with image observation could be found from: https://au.mathworks.com/help/deeplearning/ug/train-ddpg-agent-to-swing-up-and-sbalance-pendulum-with-image-observation.html.

The screen shots are shown in Fig. 5.1, Fig. 5.1(a–c) display the pendulum in various positions amid the swing.

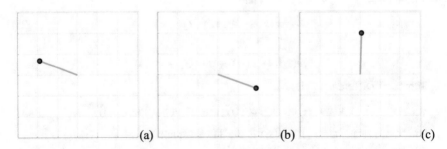

Fig. 5.1 An example shows to swing up and balance a pendulum with an image observation modelled in MATLAB

5.4 Optimization

Optimization is the core technology in deep neural networks. The optimization includes linear programming-based, nonlinear-based, dynamic-based, or neural network-based one, etc. Local optimization and global optimization are the main problems that the optimization aims to solve. The local minimum and global minimum are always sought in the optimization algorithms.

There are two categories of optimizations: unconstrained optimization and constrained optimization. Unconstrained optimization refers to the optimization without constraint conditions. Meanwhile, the constrained optimization refers to the optimization having constraint conditions. Most of the optimization problems with constraint conditions, therefore, it is constrained optimization. The constraints are usually with regard to (i.e., $w.r.t.$) or subject to (i.e., $s.t.$) constraint conditions.

Linear programming problem is

$$\mathbf{x}^* = \arg\max_{\mathbf{x}} f(\mathbf{x}), \tag{5.32}$$

it is subject to ($s.t.$) a condition $\mathbf{Ax} = \mathbf{b}$.

In linear programming, if we change the parameters such as $\mathbf{A}, \mathbf{b}$ will be changed to $\mathbf{A} + \Delta\mathbf{A}$ and $\mathbf{b} + \Delta\mathbf{b}$, where $\Delta\mathbf{A}$ and $\Delta\mathbf{b}$ are the small changes, we need to check how they will affect our optimization, find out whether the solution of this optimization problem is under control or not.

In optimization, we have multiple objective programming problem. How to find the best solution for this multiple objective programming problem is a key issue in mathematical optimization. Usually, we need to seek the derivatives. Sometimes, if we could not find the derivatives of a function for seeking the local optimization solution, we may extend the problem by using mathematical regularization [5].

For dynamic optimization, we also need to calculate the derivatives. If the derivatives could not be found, one of the solutions is to utilize genetic algorithm (GA). Modern optimization refers to nature-inspired computing. Usually, the modern optimization algorithms include genetic algorithm (GA), simulated annealing, particle swarm optimization, ant colony optimization, etc. [6].

5.5 Data Fitting

If $y = f(z, x_1, \ldots, x_n)$, $y_k = f(z_k, x_1, \ldots, x_n)$, $k = 1, \ldots, m$, $m > n$, the best solution is to minimize

$$\varepsilon(x_1, \ldots, x_n) = \sum_{i=1}^{m} (y_i - f(z_i, x_1, \ldots, x_n))^2 \tag{5.33}$$

or

$$\varepsilon(x_1, \ldots, x_n) = \sum_{i=1}^{m} (y_i - f_i(x_1, \ldots, x_n))^2. \tag{5.34}$$

Therefore,

$$\frac{\partial \varepsilon(x_1, \ldots, x_n)}{\partial x_i} = \frac{\partial}{\partial x_i} \sum_{i=1}^{m} (y_i - f_i(x_1, \ldots, x_n))^2. \tag{5.35}$$

Linear least squares problem: If the functions $f_k(x_1, \ldots, x_n)$, $k = 1, \ldots, m$ are linear, let

$$\|\mathbf{y} - A\mathbf{x}\|^2 = (\mathbf{y} - A\mathbf{x})^\tau (\mathbf{y} - A\mathbf{x})$$

be minimized as $\mathbf{x} = (x_1, \ldots, x_n)^\tau$, namely,

$$\min_{\mathbf{x} \in \mathbf{R}^n} \|\mathbf{y} - A\mathbf{x}\|$$

$$\Rightarrow \nabla_{\mathbf{x}}[(A\mathbf{x} - \mathbf{y})^\tau (A\mathbf{x} - \mathbf{y})] = 2A^\tau A\mathbf{x} - 2A^\tau \mathbf{y} = 0$$

$$\Rightarrow A^\tau A\mathbf{x} - A^\tau \mathbf{y} = 0$$

$$\Rightarrow \mathbf{x} = (A^\tau A)^{-1} A^\tau \mathbf{y}. \tag{5.36}$$

If the function $f(\mathbf{x}) = (f_1, \ldots, f_m)^\tau$ is nonlinear, $\mathbf{y} = (y_1, \ldots, y_m)^\tau$, let $\|\mathbf{y} - f(\mathbf{x})\|^2$ be minimized as $\mathbf{x} = (x_1, \ldots, x_n)^\tau$, the Jocobian matrix is

$$\frac{\partial \mathbf{J}(\mathbf{x})}{\partial \mathbf{x}} = \begin{bmatrix} \frac{\partial f_1}{\partial x_1} & \cdots & \frac{\partial f_1}{\partial x_n} \\ \cdots & \cdots & \cdots \\ \frac{\partial f_m}{\partial x_1} & \cdots & \frac{\partial f_m}{\partial x_n} \end{bmatrix} = 0 \tag{5.37}$$

The solution $\bar{\mathbf{x}}$ of the nonlinear least squares problem satisfies

$$\|\mathbf{y} - f(\bar{\mathbf{x}})\|^2 \leq \|\mathbf{y} - f(\mathbf{x})\|^2. \tag{5.38}$$

The solution is given by using Gauss–Newton method

$$\mathbf{x}^{(i+1)} := \mathbf{x}^{(i)} - \nabla^{-1} f(\mathbf{x}^{(i)}) f(\mathbf{x})^{(i)}. \tag{5.39}$$

For nonlinear function, if

$$f(\xi) = f(\mathbf{x}_0) + f'(\mathbf{x}_0)(\xi - \mathbf{x}_0) = 0, \tag{5.40}$$

then

$$\xi = \mathbf{x}_0 - \frac{f(\mathbf{x}_0)}{f'(\mathbf{x}_0)}. \tag{5.41}$$

The generalized Newton method for solving systems of equations is given by

$$\mathbf{x}_{i+1} = \mathbf{x}_i - \frac{f(\mathbf{x}_i)}{f'(\mathbf{x}_i)} \tag{5.42}$$

where $i = 0, 1, 2, \ldots$.

A sequence $\mathbf{x}_i \in \mathbf{R}^n$ is convergent if and only if for each $\varepsilon > 0$, there exists an $N(\varepsilon)$, such that $|x_l - x_m| < \varepsilon$, $\forall l, m \geq N(\varepsilon)$.

Theorem 5.1 *General convergence theorem: Let function* $y = \Phi(x)$, $x, y \in \mathbf{R}^n$ *have a fixed point* $\xi = \Phi(\xi)$ *and* $S_r(\xi) = \{x : \|x - \xi\| < r\}$ *be a neighborhood of* ξ *such as* $\Phi(\cdot)$ *is a contractive mapping in* $S_r(\xi)$, *namely,*

$$\|\Phi(x) - \Phi(y)\| \le K\|x - y\|, \tag{5.43}$$

where $K \in [0, 1]$, $x, y \in S_r(\xi)$.

For the generated sequence $x_i = \Phi(x_i)$, $i = 0, 1, 2, \ldots$, $x_i \in S_r(\xi)$,

$$\|x_{i+1} - \xi\| \le K\|x_i - \xi\|. \tag{5.44}$$

If function $y = f(x)$, $x \in S_r(x_0) = \{x : \|x - x_0\| < r\}$ has the properties:

- $\|f'(x) - f'(y)\| < \gamma\|x - y\|$, $\forall x, y \in S_r(x_0)$, $\gamma \in [0, 1]$.
- $f'(x)^{-1}$ exists, $\|f'(x)^{-1}\| < \beta$, $\forall x \in S_r(x_0)$, $\beta \in [0, 1]$.
- $\|f'(x_0)^{-1} f(x_0)\| < \alpha$, $\alpha \in [0, 1]$,

then

-

$$x_{i+1} := x_i - f'(x_i)^{-1} f(x_i), \tag{5.45}$$

where $x_i \in S_r(x_0)$, $i = 0, 1, \ldots$

-

$$\lim_{k \to \infty} x_k = \xi, \tag{5.46}$$

where $\xi \in S_r(x_0)$, $f(\xi) = 0$.
- $\forall k \ge 0$,

$$\|x_k - \xi\| < \eta \cdot \frac{h^{2k-1}}{1 - h^{2k}}, \tag{5.47}$$

where $\eta \in [0, 1]$.

Given a matrix $\mathbf{A} = (a_{ij})_{n \times n}$, find $\lambda \in \mathbf{C}$, such as the linear system of equations has a nontrivial solution $x \ne 0$.

$$(\mathbf{A} - \lambda \mathbf{I})x = 0, \tag{5.48}$$

where λ is an eigenvalue of the matrix $\mathbf{A}$, x is an eigenvector of matrix $\mathbf{A}$ associated with eigenvalue λ, the set of all eigenvalues are called the spectrum of $\mathbf{A}$.

$$\phi(\mu) = det(\mathbf{A} - \mu \mathbf{I}) \tag{5.49}$$

is called the characteristic polynomial

$$\phi(\mu) = (\mu - \lambda_1)^{\sigma_1}(\mu - \lambda_2)^{\sigma_2} \cdots (\mu - \lambda_k)^{\sigma_k}, \tag{5.50}$$

where $\sigma_i = \sigma(\lambda_i)$, $\sigma_1 + \sigma_2 + \cdots + \sigma_k = n$. Especially,

$$\phi(\mathbf{A}) = 0. \tag{5.51}$$

Given matrices $\mathbf{A}$ and $\mathbf{B}$, there exists a vector $\mathbf{x} \neq 0$,

$$\mathbf{A}\mathbf{x} = \mathbf{B}\lambda\mathbf{x}. \tag{5.52}$$

If $|\mathbf{B}| \neq 0$,

$$\mathbf{B}^{-1}\mathbf{A}\mathbf{x} = \lambda\mathbf{x} \tag{5.53}$$

and $\mathbf{A}\mathbf{x} = \lambda\mathbf{x}$, then

$$|\lambda| \leq \frac{\|\mathbf{A}\mathbf{x}\|}{\|\mathbf{x}\|}, \tag{5.54}$$

where $\rho = \max_{1 \leq i \leq n}(|\lambda_i|)$ is the spectrum radius of $\mathbf{A}$.

5.6   Questions

Question 1. How to understand the relationship among supervised learning, unsupervised learning, and reinforcement learning?

Question 2. What's the Markov decision process? Why is reinforcement learning related to Markov decision process?

Question 3. Please explain what exploration and exploitation are? How does reinforcement learning seek the tradeoff between exploration and exploitation?

Question 4. What's the Bellman equation? What are the relevant concepts in the Bellman equations?

Question 5. What's an episode in reinforcement learning?

Question 6. How is reinforcement learning related to gradient-descent methods?

Question 7. What are the Q-learning and the double Q-learning used for?

Question 8. What's linear programming? What's nonlinear programming? What's dynamic programming?

Question 9. What are modern optimization algorithms? Please list three of them.

Question 10. What's the generalized Newton method for solving systems of equations?

References

1. Littman M (2015) Reinforcement learning improves behaviour from evaluative feedback. Nature 521:445–451
2. Mnih V et al (2015) Human-level control through deep reinforcement learning. Nature 518:529–533
3. Alpaydin E (2009) Introduction to machine learning. MIT Press, Cambridge
4. Koller D, Friedman N (2009) Probabilistic graphical models. MIT Press, Cambridge
5. Goodfellow I, Bengio Y, Courville A (2016) Deep learning. MIT Press, Cambridge
6. Rao S (2009) Engineering optimization: theory and practice, 4th edn. ISBN: 978-0-470-18352-6

CapsNet and Manifold Learning

6

6.1 CapsNet

A dynamic routing mechanism for capsule networks (CapsNets) was introduced by Hinton and his team in 2017 [1]. A capsule is a set of neurons that individually activate for various properties of a visual object. A CapsNet was used to better model hierarchical relationships which is able to delineate the "Picasso problem", namely, the images that have all the right parts, are not in the correct spatial relationships. The output of a CapsNet is a vector consisting of the probability of observation, i.e., pose (e.g., position, size, orientation), deformation, velocity, etc. CapsNets replace the scalar output with vector-output capsules. Because each capsule is independent, when multiple capsules agree, the probability of correct detection or confidence is much higher.

The output of a capsule is updated by

$$b_{ij} \leftarrow b_{ij} + \hat{\mathbf{u}}_{j|i} \cdot \mathbf{v}_j, \tag{6.1}$$

where b_{ij} refers to the prior probability that capsule i in layer l should connect to capsule j in layer $l + 1$.

$$\hat{\mathbf{u}}_{j|i} = \mathbf{W}_{ij}\mathbf{u}_i, \tag{6.2}$$

where $\mathbf{W}_{ij}$ is a weight matrix. The pose vector $\mathbf{u}_i$ is rotated and translated by using $\mathbf{W}_{ij}$ into a vector $\hat{\mathbf{u}}_j$ that predicts the output of the parent capsule.

In CapsNets, the squashing function is

$$\mathbf{v}_j(s_j) = \frac{\|\mathbf{s}_j\|^2}{1 + \|\mathbf{s}_j\|^2} \frac{\mathbf{s}_j}{\|\mathbf{s}_j\|}, \tag{6.3}$$

W. Q. Yan, *Computational Methods for Deep Learning*, Texts in Computer Science, https://doi.org/10.1007/978-3-030-61081-4_6

where $\mathbf{v}_j$ is the vector output of capsule j. Capsules $\mathbf{s}_j$ in the next layer are fed from the sum of the predictions from all capsules in the previous layers with a coupling coefficient c_{ij},

$$\mathbf{s}_j = \sum_i c_{ij} \hat{\mathbf{u}}_{j|i} \qquad (6.4)$$

and

$$c_{ij} = softmax(\mathbf{b}_i) = \frac{\exp(b_{ij})}{\sum_k \exp(b_{ik})}, \qquad (6.5)$$

where c_{ij} is coupling coefficients, b_{ij} is the log prior probability, initially, $b_{ij} := 0$. Eventually, the network is trained by minimizing the loss function

$$L_k = T_k \max(0, m^+ - \|\mathbf{v}_k\|)^2 + \lambda(1 - T_k) \max(0, \|\mathbf{v}_k\| - m^-)^2, \qquad (6.6)$$

where $m^+ = 0.9$, $m^- = 0.1$, and $\lambda = 0.5$,

$$T_k = \begin{cases} 1 & \text{digit of class } k \text{ present} \\ 0 & \text{others.} \end{cases}$$

CapsNets have multiple conceptual advantages, which learn topological relationship, the networks are organized in a hierarchical way. CapsNets have the attribute with viewpoint invariance and better generalization to new viewpoints. Moreover, CapsNets have been applied to image segmentation, which works like SegNets and U-Nets. The two deep learning networks were designed specifically for image segmentation.

U-Nets [2, 3] were also applied to pixel-wise regression and small-size object detection and recognition. U-Net is a convolutional neural network that consists of a contracting path and an expansive path. The contracting path follows the typical architecture of a convolutional network. The network has U-shaped architecture, consisting of repeated applications of convolution, followed by a rectified linear unit (ReLU) and a max pooling operation. The design was based on fully convolutional network (FCN) and its architecture was modified and extended with fewer training images and to yield more precise segmentation. U-Net uses a pixel-wise softmax cross-entropy as the loss function. The softmax function is defined as

$$p_k(\mathbf{p}) = \frac{\exp(a_k(\mathbf{p}))}{\sum \exp(a_k(\mathbf{p}))}, \qquad (6.7)$$

where $p_k(\mathbf{p})$ is the softmax function at pixel $\mathbf{p}$ within the feature channel k, $k = 1, 2, \ldots, K$, K is the number of classes; $a_k(\mathbf{p})$ is the activation function at the pixel position $\mathbf{p} = (x, y) \in \Omega = [a, b] \times [c, d] \subset \mathscr{R}^2$, $x \in [a, b]$ and $y \in [c, d]$ are the intervals of the image region in horizontal and vertical directions, respectively.

The pixel-wise softmax cross-entropy is

$$L(p) = \sum_{\mathbf{p} \in \Omega} w(\mathbf{p}) \log p(a_l(\mathbf{p})), \qquad (6.8)$$

where $a_l(\mathbf{p})$ is the softmax function with a channel label l, $0 < l \le K$ at pixel $\mathbf{p}$; $w(\mathbf{p})$ is a weight map at pixel $\mathbf{p}$, the map with a pixel-wise loss weight forces the U-Net network to learn from the border pixels. The weight map is computed as

$$w(\mathbf{p}) = w_c(\mathbf{p}) + w_0(\mathbf{p}) \exp\left(-\frac{(d_1(\mathbf{p}) + d_2(\mathbf{p}))^2}{2\sigma^2}\right), \tag{6.9}$$

where $w_c(\mathbf{p})$, $w_0(\mathbf{p})$ and $\sigma \ne 0$ are parameters, which are treated as constants. $d_1(\mathbf{p})$ and $d_2(\mathbf{p})$ are the first longest distance and the second longest distance from pixel $\mathbf{p}$ to its border pixels.

Meanwhile, SegNet [4] uses all of the pretrained convolutional layer weights from VGG [5] neural networks as pretrained weights, which were developed by the University of Cambridge, UK. The encoder network consists of 13 convolutional layers which correspond to the first 13 convolutional layers in the VGG16 network. Each encoder layer has a corresponding decoder layer, hence, the decoder network has 13 layers. The final decoder output is fed to a multiclass softmax classifier to produce class probabilities for each pixel independently. The cross-entropy loss was used as the objective function for training SegNet, the loss is summed up over all the pixels in a mini-batch. SegNet only stores the max pooling indices of the feature maps and uses them in its decoder network to achieve good performance.

Moreover, SegNet [4] is a deep encoder–decoder architecture for multiclass pixel-wise segmentation. SegNet is effective for a real-time urban road scene segmentation as well as indoor scene understanding. The architecture consists of a sequence of non-linear processing layers (encoders) and a corresponding set of decoders followed by a pixelwise classifier. Typically, each encoder consists of one or more convolutional layers with batch normalization and a ReLU nonlinearity, followed by nonoverlapping max pooling and subsampling. One key component of the SegNet is to perform upsampling (bilinear interpolation) in the decoders for low-resolution feature maps. The entire architecture can be trained by using stochastic gradient descent.

Downsampling is implemented by using pooling operations such as max pooling, average pooling, etc. Pertaining to upsampling, the nearest neighbour, bilinear, and bicubic interpolation methods [6] are employed to the operation. For a bilinear interpolation, we have a region $\Omega = [a, b] \times [c, d]$ and a region $\Omega' = [a', b'] \times [c', d']$ in images I and I', respectively, the parameters $s \in [0, 1]$ and $t \in [0, 1]$ establish the mapping relationship between region $\Omega \subset I$ to $\Omega' \subset I'$. That means, given pixel $\mathbf{p} \in \Omega \subset I$, we will get the corresponding pixel $\mathbf{p}' \in \Omega' \subset I'$ via parameters $s_0, t_0 \in [0, 1]$. Hence $\mathbf{p}(s, t) = \mathbf{p}'(s, t)$, $\forall s, t \in [0, 1]$, namely,

$$\mathbf{p}_A = \mathbf{p}(0, 0) = \mathbf{p}'(0, 0) = \mathbf{p}'_{A'}, \tag{6.10}$$

$$\mathbf{p}_B = \mathbf{p}(1, 0) = \mathbf{p}'(1, 0) = \mathbf{p}'_{B'}, \tag{6.11}$$

$$\mathbf{p}_C = \mathbf{p}(0, 1) = \mathbf{p}'(0, 1) = \mathbf{p}'_{C'}, \tag{6.12}$$

and

$$\mathbf{p}_D = \mathbf{p}(1, 1) = \mathbf{p}'(1, 1) = \mathbf{p}'_{D'}, \tag{6.13}$$

where pixels at the four corners points $\mathbf{p}_A, \mathbf{p}_B, \mathbf{p}_C, \mathbf{p}_D \in \Omega$ correspond to the four corners points at $\mathbf{p}_{A'}, \mathbf{p}_{B'}, \mathbf{p}_{C'}, \mathbf{p}_{D'} \in \Omega'$. Thus,

$$\mathbf{p}'(s_0, t_0)$$
$$= t_0 \cdot [s_0 \cdot \mathbf{p}'(0, 0) + (1.0 - s_0) \cdot \mathbf{p}'(1, 0)] + (1.0 - t_0) \cdot [s_0 \cdot \mathbf{p}'(1, 0) + (1.0 - s_0) \cdot \mathbf{p}'(1, 1)].$$
$$(6.14)$$

In the matrix form,

$$\mathbf{p}'(s_0, t_0) = (t_0, 1.0 - t_0)\mathbf{M}'(s_0, 1 - s_0)^\tau \tag{6.15}$$

where

$$\mathbf{M}' = \begin{bmatrix} \mathbf{p}'(0, 0) \ \mathbf{p}'(1, 0) \\ \mathbf{p}'(0, 1) \ \mathbf{p}'(1, 1) \end{bmatrix}. \tag{6.16}$$

Meanwhile,

$$\mathbf{p}(s_0, t_0)$$
$$= t_0 \cdot [s_0 \cdot \mathbf{p}(0, 0) + (1.0 - s_0) \cdot \mathbf{p}(1, 0)] + (1.0 - t_0) \cdot [s_0 \cdot \mathbf{p}(1, 0) + (1.0 - s_0) \cdot \mathbf{p}(1, 1)]. \tag{6.17}$$

Similarly, in the matrix form,

$$\mathbf{p}(s_0, t_0) = (t_0, 1.0 - t_0)\mathbf{M}(s_0, 1 - s_0)^\tau, \tag{6.18}$$

where

$$\mathbf{M} = \begin{bmatrix} \mathbf{p}(0, 0) \ \mathbf{p}(1, 0) \\ \mathbf{p}(0, 1) \ \mathbf{p}(1, 1) \end{bmatrix} \tag{6.19}$$

6.2 Manifold Learning

Manifold [7–9] is a generalization of curve and surface, a line is 1D manifold, a curve is 2D manifold, a surface is 3D manifold, n-dimensional manifold is n-manifold. In manifold, *chart* is an important concept in Euclidean space, which is related to the neighbourhood. Atlas is a local Euclidean space, a collection of topology with continuity of infinity. If a manifold is smooth, we call it as smooth manifold. Manifold includes analytic manifold, complex manifold, Euclidean manifold, topological manifold, etc.

In manifold, if the left continuity of function $f(x) \in [a, b]$, $x \in \mathscr{R}$ is equal to the right continuity, namely,

$$\lim_{x \to x_0} f(x) = \lim_{x \to x_0^+} f(x) = f(x_0^+) = f(x_0^-) = \lim_{x \to x_0^-} f(x), \tag{6.20}$$

where C^k is continuous if $f^k(x)$ with derivatives of order $k \in \mathscr{Z}^+$. C^∞ is continuous with k tends to infinity, i.e., $k \to \infty$.

Topological manifold is defined in Hausdorff space with Euclidean distance. Topological space has a base, n-manifold has the base with a dimension n, which is countable; thus, the topological manifold is countable.

Two charts have a compatible relationship, if we have invertible functions $\Psi^{-1}(\cdot)$ and $\Phi^{-1}(\cdot)$, $x \in C_1$ and $y \in C_2$, $y = \Phi(x)$, $\Phi^{-1}\Phi(x) = x$, $\Psi^{-1}\Psi(x) = x$, $\Psi^{-1}\Phi\Psi(x) = \Phi(x)$ and $\Phi^{-1}\Psi\Phi(x) = \Psi(x)$.

Homomorphism is a mapping which keeps the relationship

$$\Phi(u \cdot v) = \Phi(u) \circ \Phi(v), \forall u, v \in \mathbf{A}. \tag{6.21}$$

Before and after the mapping, the defined operations are kept.

Manifold is homomorphism, that means the manifold is defined on a continuous domain. Riemann manifold is a smooth manifold based on derivatives or tangent vectors. The centre of manifold is defined as a set $\mathbf{C} = \{x : ax = xa = 0, x \in \mathbf{K}, a \in \mathbf{A}\}$.

Based on mapping of manifold, if we have two functions $f(\cdot)$ and $g(\cdot)$, all are C^∞, then the composite function $f \circ g$ will be continuous infinity C^∞, i.e., $f \circ g \in C^\infty$.

Manifold learning has been applied to medical image processing, data compression, data dimensionality reduction, noise removal, etc. PCA (Principal Component Analysis) is a linear dimensionality reduction method, while manifold learning is for nonlinear dimensionality reduction which could be applied to noise removal. The data dimensionality reduction has been applied to resolve the "curse of dimensionality" problem.

PCA (principal component analysis) refers to an orthogonal linear transformation that transforms the given data to a new coordinate system so that the greatest variance by using scalar projection of the given data replies on the principal components.

The principal component decomposition of a raw vector $\mathbf{x}$ is given as

$$\mathbf{b} = \mathbf{x}\mathbf{A}, \tag{6.22}$$

where $\mathbf{A} = (a_{ij})_{p \times p}$ is a weight matrix, $p \in \mathcal{Z}$, $a_{ij} \in \mathcal{R}$. The transformation maps a data vector $\mathbf{x} = (x_{ij})_{1 \times p}$ from an original space to a vector $\mathbf{b} = (b_{ij})_{1 \times p}$ in new space, $x_{ij}, b_{ij} \in \mathcal{R}$. If we have $0 < l < p, l, p \in \mathcal{Z}$, we have another mapping

$$\mathbf{b}_l = \mathbf{x}\mathbf{A}_l, \tag{6.23}$$

where $\mathbf{A}_l = (a_{ij})_{p \times l}$, $\mathbf{b}_l = (b_{ij})_{1 \times l}$, $a_{ij}, x_{ij}, b_{ij} \in \mathcal{R}$. We expect to minimize the squared reconstruction error

$$\varepsilon = \|\mathbf{x} - \mathbf{x}_l\|_2^2 > 0, \tag{6.24}$$

where $\mathbf{x}_l$ has a lower dimension than the vector $\mathbf{x}$, namely, $l < p$. Hence, the principal components of the original data are preserved, and the dimension of the raw data is reduced.

In order to implement dimensionality reduction by using PCA through the covariance method, we seek the eigenvectors and eigenvalues of the covariance matrix

$$\mathbf{B} = \mathbf{X} - \mathbf{h}\mathbf{u}^\tau, \tag{6.25}$$

where $\mathbf{X} = (x_{ij})_{n \times p}$ is a matrix consisting of n given vectors $\mathbf{x}_i = (x_{ij})_{1 \times p}$, $\mathbf{h} = (h_{ij})_{1 \times n}$, $h_{ij} = 1$ and $\mathbf{u} = (u_{ij})_{1 \times p}$,

$$u_{1,j} = \frac{1}{n} \sum_{i=1}^{n} x_{i,j}, \tag{6.26}$$

where $j = 1, 2, \ldots, p, i = 1, 2, \ldots, n, x_{ij}, u_{i,j} \in \mathscr{R}$. Thus, we have the covariance matrix $\mathbf{C}$ from matrix $\mathbf{B}$

$$\mathbf{C} = \frac{1}{n-1} \mathbf{B}^* \mathbf{B}, \tag{6.27}$$

where '*' is the conjugate transpose operator. Thus,

$$\mathbf{V}^{-1} \mathbf{C} \mathbf{V} = \Lambda = diag(\lambda_1, \lambda_2, \ldots, \lambda_n), \tag{6.28}$$

where $\mathbf{V}$ is the matrix consisting of eigenvectors, Λ is the diagonal matrix of eigenvalues of $\mathbf{C}$, $\lambda_i \neq 0, i = 1, 2, \ldots, n$ are eigenvalues. Namely, the eigenvalues satisfy the characteristic polynomial

$$f(\lambda) = |\mathbf{C} - \lambda \mathbf{I}| = 0, \tag{6.29}$$

where $\mathbf{I}$ is the identity matrix. We sort the eigenvalues in decreasing order; thus, the order of the corresponding eigenvectors is swapped. We thus select the principal components using Eq. (6.30)

$$\delta = \frac{\sum_{i=1}^{l} |\lambda_i|}{\sum_{i=1}^{p} |\lambda_i|} > 0, \ (p \geq l > 0). \tag{6.30}$$

If $\delta > 0.90$, for example, $\lambda_i, i = 1, 2, \ldots, l \ (p \geq l > 0)$ are the main components of matrix $\mathbf{C}$.

$$\mathbf{V}_l^{-1} \mathbf{C}_l \mathbf{V}_l = \Lambda = diag(\lambda_1, \lambda_2, \ldots, \lambda_l), \tag{6.31}$$

where $\mathbf{V}_l$ is the matrix consisting of the eigenvectors corresponding to eigenvalues $\lambda_1, \lambda_2, \ldots, \lambda_l$. Correspondingly, we have found $l \in \mathscr{Z}^+$, which satisfies $\mathbf{b}_i = \mathbf{x} \mathbf{C}_l$, $\forall \mathbf{x}$.

In manifold learning, we always assume that we can always find the lower dimensional data using dimensionality reduction. We assume the lowest dimensional data is hidden or embedded in noisy data. This is our base point or assumption of using manifolds to machine learning, or directly we call it as manifold learning.

In manifold learning, we construct the relationship matrix between the nodes or vertices x_i and $x_j, i, j = 1, 2, \ldots, n$ of a given graph G, $\mathbf{W} = \{w_{ij}\}$. Given a dataset with training samples $\mathbf{X} = \{\mathbf{x}_i\}_{i=1}^{n}, \mathbf{x}_i \in \mathbf{R}^d$, $G = <\mathbf{X}, \mathbf{W}>$ is an undirected graph, $\mathbf{W} = (w_{ij})_{n \times n}$ is the similarity or affinity matrix, $w_{ij} \in [0, 1]$

$$w_{ij} = \exp \left(-\frac{\|\mathbf{x}_i - \mathbf{x}_j\|^2}{\gamma_i \gamma_j} \right), \tag{6.32}$$

where $\gamma_i = \|\mathbf{x}_i - \mathbf{x}_i\|$ is the local scale of data samples in the neighbourhood of $\mathbf{x}_i$. $\mathbf{x}_i$ is the k-nearest of $\mathbf{x}_i$.

Graph Laplacian is

$$\mathbf{L} = \mathbf{D} - \mathbf{W}, \tag{6.33}$$

where $\mathbf{D} = (d_{ij})_{n \times n}$, $d_{ii} = \sum_j w_{ij}, \forall i$.

For dimensionality reduction of data samples $\{\mathbf{y}_i\}_{i=1}^n$, $\mathbf{y}_i \in \mathbf{R}^D$, $D \gg d$, the eigenspectrum-based method is

$$\mathbf{Ly} = \lambda \mathbf{By}, \tag{6.34}$$

where $\mathbf{yBy}^\tau = \mathbf{I}$, $\mathbf{I}$ is the identify matrix. Thus, we get $\mathbf{y}^*$ by using

$$\mathbf{y}^* = \arg \min_{\mathbf{yBy}^\tau = \mathbf{I}} \mathbf{yLy}^\tau. \tag{6.35}$$

Given a graph, a matrix of edge weights $\mathbf{x}_i \in \mathcal{R}^d$, $\mathbf{W} = (w_{ij})_{n \times n}$,

$$\mathbf{L} = \mathbf{D} - \mathbf{W}, \tag{6.36}$$

where $\mathbf{D} = (d_{ii})_{n \times n}$, $d_{ii} = \sum_j w_{ij}$,

$$w_{ij} = \begin{cases} \exp\left(-\frac{\|\mathbf{x}_i - \mathbf{x}_j\|^2}{2\sigma^2}\right) & x_j \in \mathbf{N}(i) \\ 0 & \text{Others,} \end{cases} \tag{6.37}$$

where w_{ij} is the Gaussian kernal, $\mathbf{N}(i)$ is the neighbourhood of x_i.

$$\mathbf{Ly} = \lambda \mathbf{Dy}, \tag{6.38}$$

where $\mathbf{Y} = (\mathbf{y}_i)_n$ is the output.

An example of manifold learning is available from the site: https://scikit-learn.org. Python has been applied to implement the methods, the examples are applied to reduce the data dimension of a Swiss roll (Fig. 6.1).

A MATLAB example is available to demonstrate the Laplacian eigenmap in manifold learning at: https://www.mathworks.com/matlabcentral/fileexchange/36141-laplacian-eigenmap-diffusion-map-manifold-learning. The result is shown in Fig. 6.2 for the purpose of recovering low-dimensional geometries.

6.3 Questions

Question 1. What is the difference between CNNs and CapsNets? What can CapsNets bring to us?

Question 2. What is the loss function of CapsNets? What is the squashing function of CapsNets?

Question 3. What are the three key concepts in manifold?

Question 4. Please list the algorithms for data dimensionality (dimension) reduction and give the differences between them.

Question 5. Why could manifold learning be applied to dimensionality reduction?

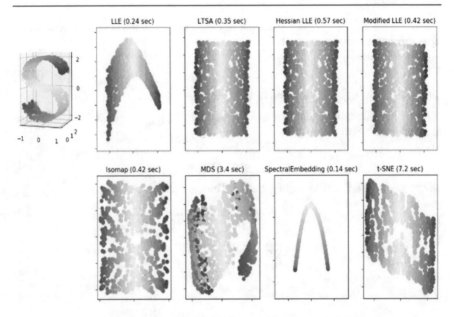

Fig. 6.1 Python for dimensionality reduction of Swiss roll by using manifold learning

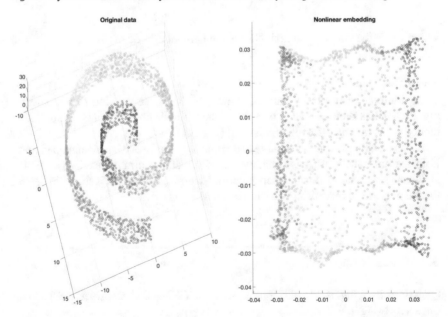

Fig. 6.2 MATLAB for the Laplacian eigenmap method of the Swiss roll by using manifold learning

References

1. Sabour S, Frosst N, Geoffrey E (2017) Hinton dynamic routing between capsules. In: The conference on neural information processing systems (NIPS), USA
2. Yao W, Zeng Z, Lian C, Tang H (2018) Pixel-wise regression using U-Net and its application on pansharpening. Neurocomputing 312:364–371
3. Ronneberger O, Fischer P, Brox T (2015) U-Net: convolutional networks for biomedical image segmentation. In: International conference on medical image computing and computer-assisted intervention. Springer, pp 234–241
4. Badrinarayanan V, Handa A, Cipolla R (2017) SegNet: a deep convolutional encoder-decoder architecture for robust semantic pixel-wise labelling. IEEE Trans Pattern Anal Mach Intell 39(12):2481–2495
5. Simonyan K, Zisserman A (2015) Very deep convolutional networks for large-scale image recognition. In: International conference on learning representations
6. Keys R (1981) Cubic convolution interpolation for digital image processing. IEEE Trans Acoust Speech Signal Process 29(6):1153–1160
7. Zhu B et al (2018) Image reconstruction by domain-transform manifold learning. Nature 555:487–492
8. Zheng N, Xue J (2009) Statistical learning and pattern analysis for image and video processing. Springer, Berlin
9. Tu L (2011) Introduction to manifold, 2nd edn

Boltzmann Machines

7.1 Boltzmann Machine

Hopfield networks is a typical neural network. With regard to this network, on an equidistance circle, we link all the nodes together. Boltzmann machines [1] are seen as the stochastic and generative counterpart of Hopfield nets.

Boltzmann machine is a general "connectionist" approach to learn arbitrary probability distributions over binary vectors. For a d-dimensional binary random vector $\mathbf{x} \in \{0, 1\}^d$, an energy-based model is

$$P(\mathbf{x}) = \frac{\exp(-E(\mathbf{x}))}{Z} \tag{7.1}$$

$$E(\mathbf{x}) = -\mathbf{x}\mathbf{U}^\top \mathbf{x} - \mathbf{b}^\top \mathbf{x}, \tag{7.2}$$

where $E(\mathbf{x})$ is the energy function, $Z(\cdot)$ is the partition function, $\sum_{\mathbf{x}} P(\mathbf{x}) = 1$, U is the weight parameters, $\mathbf{x}$ is bias parameters. Units $\mathbf{x}$ could be decomposed into visible $\mathbf{v}$(visible) and hidden (latent) $\mathbf{h}$ ones

$$E(\mathbf{v}, \mathbf{h}) = -\mathbf{v}^T \mathbf{R}\mathbf{v} - \mathbf{v}^T \mathbf{W}\mathbf{h} - \mathbf{h}^T \mathbf{S}\mathbf{h} - \mathbf{b}^T \mathbf{v} - \mathbf{c}^T \mathbf{h}. \tag{7.3}$$

7.2 Restricted Boltzmann Machine

Restricted Boltzmann Machines (RBM) [2] further restrict Boltzmann machines to those without visible–visible and hidden–hidden connections. A deep Boltzmann machine (DBM) is a type of binary pairwise MRF (undirected probabilistic graphical model) with multiple layers of hidden random variables [3, 4].

W. Q. Yan, *Computational Methods for Deep Learning*, Texts in Computer Science, https://doi.org/10.1007/978-3-030-61081-4_7

The global energy E in a Boltzmann machine is identical in the form to that of a Hopfield network. Mathematically,

$$E \stackrel{\Delta}{=} - \left(\sum_{i<j} w_{ij} s_i \cdot s_j + \sum_i \theta_i \cdot s_j \right) \tag{7.4}$$

where

- w_{ij} is the connection strength between unit j and unit i.
- s_i is the state of unit i, $s_i \in \{0, 1\}$.
- θ_i is the bias of unit i in the global energy function.
- w_{ij} is represented as a symmetric matrix $\mathbf{W} = (w_{ij})_{N \times N}$, with zeros along the diagonal.
- The probability of the ith unit is

$$p_i \stackrel{\Delta}{=} \frac{1}{1 + \exp\left(-\frac{E_i}{T}\right)}. \tag{7.5}$$

RBM [2] is intractable and bipartite graph, its energy function is

$$E(\mathbf{v}, \mathbf{h}) = -\mathbf{v}^\top \mathbf{W} \mathbf{h} - \mathbf{b}^\top \mathbf{v} - \mathbf{c}^\top \mathbf{h}. \tag{7.6}$$

Therefore,

$$P(\mathbf{v} = v, \mathbf{h} = h) = \frac{\exp(-E(\mathbf{v}, \mathbf{h}))}{Z} \tag{7.7}$$

and

$$Z = \sum_{\mathbf{v}} \sum_{\mathbf{h}} \exp(-E(\mathbf{v}, \mathbf{h})). \tag{7.8}$$

A restricted Boltzmann machine (RBM) [2] is a generative neural network that can learn a probability distribution from its set of inputs. The energy function from product of expert (POE) [5] is

$$E(v, h) = - \sum_{ij} w_{ij} h_i v_j - \sum_j b_j v_j + \sum_i c_i h_i. \tag{7.9}$$

The probabilities are

$$p(v) \stackrel{\Delta}{=} \frac{\sum_h e^{-E(v,h)}}{\sum_{v,h} e^{-E(v,h)}}, \tag{7.10}$$

$$p(h) \stackrel{\Delta}{=} \frac{\sum_v e^{-E(v,h)}}{\sum_{v,h} e^{-E(v,h)}}, \tag{7.11}$$

and

$$p(v, h) \stackrel{\Delta}{=} \frac{e^{-E(v,h)}}{\sum_{v,h} e^{-E(v,h)}}; \tag{7.12}$$

hence,

$$p(v|h) = \frac{e^{-E(v,h)}}{\sum_v e^{-E(v,h)}}. \tag{7.13}$$

The loss function is

$$L(\theta) = \prod_v L(\theta|v) = \prod_v p(v), \theta = (\mathbf{W}, \mathbf{b}, \mathbf{c}). \tag{7.14}$$

The derivatives are

$$\frac{\partial L(\theta)}{\partial \theta} = \sum_v \frac{\partial \ln L(\theta|v)}{\partial \theta} = \sum_v \frac{\partial \ln p(v)}{\partial \theta}, \tag{7.15}$$

$$\ln p(v) = \ln \left(\sum_h e^{-E(v,h)} \right) - \ln \left(\sum_{v,h} e^{-E(v,h)} \right), \tag{7.16}$$

and

$$\frac{\partial L(\theta)}{\partial \theta} = E_{p(h|v)} \left(-\frac{\partial E(v,h)}{\partial \theta} \right) - E_{p(h,v)} \left(-\frac{\partial E(v,h)}{\partial \theta} \right). \tag{7.17}$$

The energy function from product of expert (POE) [5] is

$$E(v, h) = -\sum_{ij} w_{ij} h_i v_j - \sum_j b_j v_j + \sum_i c_i h_i \tag{7.18}$$

$$\frac{\partial \ln p(v)}{\partial w_{ij}} = p(h_i = 1|v) v_j - \sum_v p(v) p(h_i = 1|v) v_j \tag{7.19}$$

$$\frac{\partial \ln p(v)}{\partial b_j} = v_j - \sum_v p(v) v_j \tag{7.20}$$

and

$$\frac{\partial \ln p(v)}{\partial c_i} = p(h_i = 1|v) - \sum_v p(v) p(h_i = 1|v). \tag{7.21}$$

A DBM is an energy-based model

$$P(v, h) = \frac{1}{Z(\theta)} \exp(-E(v, h^{(1)}, h^{(2)}, h^{(3)}, h^{(4)}; \theta)) \tag{7.22}$$

where $E(\cdot)$ is the energy function, v is visible layer, $h^{(i)}$ are hidden layers, $i = 1, 2, 3$,

$$E(v, h; \theta) = -v^\top W^{(1)} h^{(1)} - h^{(1)\top} W^{(2)} h^{(2)} - h^{(2)\top} W^{(3)} h^{(3)}. \tag{7.23}$$

In the case with two hidden layers,

$$P(v_i = 1|h^{(1)}) = \sigma(W_{i,:}^{(1)} h^{(1)}), \tag{7.24}$$

$$P(h_i^{(1)} = 1|v, h^{(2)}) = \sigma(v^\top W_{:,i}^{(1)} + W_{i,:}^{(2)} h^{(2)}), \tag{7.25}$$

and

$$P(h_k^{(2)} = 1|h^{(1)}) = \sigma(h^{(1)} W_{:,k}^{(2)}). \tag{7.26}$$

Restricted Boltzmann machines (RBMs) are created by Geoff Hinton for unsupervised learning. The aim of RBMs is to find patterns in data by reconstructing the inputs using only two layers (the visible layer and the hidden layer). The MATLAB and Python source codes for implementing RBMS all are available. An example from the site: https://scikit-learn.org/ shows how to improve the classification accuracy using a RBM.

7.3   Deep Boltzmann Machine

A deep Boltzmann machine (DBM) is a type of binary pairwise MRF with multiple layers of hidden random variables, which is a network of symmetrically coupled stochastic binary units. The probability assigned to vector $\mathbf{v}$ is

$$P(\mathbf{v}) = \frac{1}{Z} \sum_h e^{[\sum_{ij}(W_{ij}^{(1)} v_i h_j^{(1)}) + \sum_{jl}(W_{jl}^{(2)} h_j^{(1)} h_l^{(2)}) + \sum_{lm}(W_{lm}^{(3)} h_l^{(2)} h_m^{(3)})]}, \qquad (7.27)$$

where $\mathbf{h} = \{h^{(1)}, h^{(2)}, h^{(3)}\}$ is the set of hidden units, $\mathbf{W} = \{W^{(1)}, W^{(2)}, W^{(3)}\}$ is the model parameters, $p(h_i|v)$ and $p(v_i|h)$ is independent,

$$p(h|v) = \sigma \left(\sum_j w_{ij} \cdot v_j + c_i \right), \qquad (7.28)$$

where $\sigma(x) = 1/(1 + e^{-x})$ and

$$p(h|v) = \prod_i p(h_i|v), \, p(v|h) = \prod_i p(v_i|h). \qquad (7.29)$$

For DBM, if $p(h_i|v)$ and $p(v_i|h)$ are independent, then $p(h|v) = \prod_i p(h_i|v)$ or $p(v|h) = \prod_i p(v_i|h)$.

DBMs are understood as multilayer perceptron with restricted Boltzmann machines. Deep belief nets are thought as Bayesian belief nets [6, 7] with deep Boltzmann machines. In machine learning,

$$\mathbf{Y} = F(\mathbf{X}, \theta), \qquad (7.30)$$

where θ is the parameter vector, $\mathbf{X}$ is the input, $\mathbf{Y}$ is the labeled dataset.

$$E_A(\theta) = \frac{1}{N} \sum_{i=1}^{N} E(x_i, d_i, \theta), \qquad (7.31)$$

where $\{(x_i, d_i)\}$ is the training set, $x_i \in \mathbf{X}, d_i \in \mathbf{D}$.

$$\theta_{k+1} := \theta_k - \varepsilon \cdot \frac{\partial E(\theta)}{\partial \theta}, k = 1, 2, \dots. \qquad (7.32)$$

7.4   Probabilistic Graphical Models

Graphical model is everywhere, it is not deep learning, but it is much wider than deep learning model. There are two types of graphs: directed graphs and undirected graphs. Bayesian networks and HMMs are directed networks, Markov random field (MRF) is a typical undirected model. In fact, we have template-based graphical models, we need to create and fill up our content there. The template-based models are general, not specific, and concrete. Meanwhile, generative model is to create a new model, while discriminative model is to adjust or modify the model to suit our requirements. Hybrid models are to combine all of these models together.

Our variables include target variables (output), observable variables (input), and latent variables (hidden). Inferences include exact inferences and approximate inferences. Uncertainty is the typical concept in machine learning. The graphical inferences can help us from what we know to infer what we do not know. Inferences could assist us to test models and find the sensitivities and errors.

Why do we study graphical models? Graphical model is a simple way to visualize the structure of a probabilistic model and can be used to design and motivate new models [5].

In graphical model, we have various presentations, we need to find which model is the best one for solving our problem.

In graphical model, we have the parameters learning, feature learning, and knowledge learning. Latency refers to the hidden layers that we do not know.

Graphical model insights into the properties of a model, including conditional independence properties, can be obtained by inspection of the graph.

Complex computations, required to perform inference and learning in sophisticated models, can be expressed in terms of graphical manipulations, in which underlying mathematical expressions are carried out implicitly.

Bayes' theorem is the base stone of modern pattern classification [8]. From the prior probability, likelihood, and evidence infer the posterior probability for a class.

Bayesian model is a simple yet highly effective method for pattern classification in machine learning. The joint probability is

$$P(G, S, R) = P(G|S, R) \cdot P(S|R) \cdot P(R). \tag{7.33}$$

Meanwhile, the conditional probability is

$$P(G = T|R = T) = \frac{P(R = T, G = T)}{P(R = T)} = \frac{P(G = T, S, R = T)}{P(G = T, S, R)}. \tag{7.34}$$

Naive Bayesian model is a family of simple probabilistic classifiers based on applying Bayes' theorem with strong (Naive) assumptions between the features. Naive Bayesian model has been applied to classify spams using the words appeared in the emails.

Influence diagram [9, 10] is a decision-theoretic diagram framework for making decisions under uncertainty, Bayesian network is a directed acyclic graph (DAG).

Markov random field (MRF) [4] is undirected graphs with probability distributions. Factor graphs encompass both Bayesian networks and Markov networks. Factorization is a product of factors over cliques in the graph [5].

$$P(X = x) = \frac{1}{Z} \exp \sum_k \sum_{i=1}^{N_k} w_{ki} f_{ki}(x_{\{k\}}) \tag{7.35}$$

and

$$Z = \sum_{x \in \mathscr{X}} \exp \sum_k \sum_{i=1}^{N_k} w_{ki} f_{ki}(x_{\{k\}}). \tag{7.36}$$

A distribution P_Φ is a Gibbs distribution parameterized by using a set of factors $\Phi = (\phi_1(D_1), \ldots, \phi_k(D_k))$ if it is defined as

$$P_\Phi = (X_1, \ldots, X_n) = \frac{1}{Z}\phi_1(D_1) \times \cdots \times \phi_m(D_m), \tag{7.37}$$

where $Z = \sum_{X_1, X_2, \ldots, X_n} \phi_1(D_1) \times \cdots \times \phi_m(D_m)$ is a normalizing constant called partition function.

A conditional random field (CRF) is an undirected graph [11, 12], the network is annotated with a set of factors $\phi_1(D_1), \ldots, \phi_m(D_m)$, a conditional distribution is

$$P(Y|X) = \frac{1}{Z}\prod_{i=1}^{m}\phi_i(Y_i, Y_{i+1}) \tag{7.38}$$

and

$$Z = \sum_Y \prod_{i=1}^{m}\phi_i(X_i, Y_i). \tag{7.39}$$

A CRF over $X = \{X_1, X_2, \ldots, X_n\}$, $Y = \{0, 1\}$,

$$\phi_i(X_i, Y) = \exp(w_i \mathbf{I}(X_i = 1, Y = 1)) \tag{7.40}$$

and

$$P(Y = 1|x_1, \ldots, x_k) = \sigma\left(w_0 + \sum_{i=1}^{k} w_i x_i\right), \tag{7.41}$$

where $\sigma(\cdot)$ is a sigmoid function.

Logistic CPD is

$$P(Y = 1|X_1, \ldots, X_n) = \sigma\left(w_0 + \sum_{i=1}^{N} w_i\right), \tag{7.42}$$

where $\sigma(\cdot)$ is a sigmoid function. Logistic distributions only have two labels: '0' and '1'.

Linear or multivariate Gaussian distribution is

$$p(Y|\mathbf{x}) = \mathbf{N}(\beta_0 + \beta\mathbf{x}; \sigma^2). \tag{7.43}$$

Conditional Bayesian network is

$$P(Y|X) = \sum_Z P(Y, Z|X) = \prod_{X \in Y \bigcup Z} P(X|P_X). \tag{7.44}$$

Multivariate Gaussian distribution is

$$P(X) = \frac{1}{(2\pi)^{(n/2)}|\Sigma|^{1/2}}e^{(X-\mu)^\tau \Sigma^{-1}(X-\mu)}. \tag{7.45}$$

A joint normal distribution over $\{X, Y\}$ is $P(X, Y) \sim \mathbf{N}(\mu, \Sigma)$,

$$\mu_{(n+m)\times 1} = \begin{pmatrix} (\mu_X)_{n\times 1} \\ (\mu_Y)_{m\times 1} \end{pmatrix} \tag{7.46}$$

and

$$\Sigma_{(n+m)\times(n+m)} = \begin{pmatrix} (\Sigma_{XX})_{n\times n} & (\Sigma_{XY})_{n\times m} \\ (\Sigma_{YX})_{m\times n} & (\Sigma_{YY})_{m\times m} \end{pmatrix}. \tag{7.47}$$

For Gaussian Bayesian networks, if

$$p(Y|x) \sim \mathbf{N}(\beta_0 + \beta^\tau x; \sigma^2), \tag{7.48}$$

then

$$p(Y) \sim \mathcal{N}(\mu_Y; \sigma_Y^2), \tag{7.49}$$

$$\mu_Y = \beta_0 + \beta^\tau x, \tag{7.50}$$

and

$$\sigma_Y^2 = \sigma^2 + \beta^\tau \Sigma \beta. \tag{7.51}$$

The conditional density is

$$P(Y|X) \sim \mathbf{N}(\beta_0 + \beta^\tau X; \sigma^2), \tag{7.52}$$

where

$$\beta_0 = \mu_Y \Sigma_{YX} \Sigma_{XX}^{-1} \mu_X, \tag{7.53}$$

$$\beta = \Sigma_{XX}^{-1} \Sigma_{YX}, \tag{7.54}$$

and

$$\sigma^2 = \Sigma_{YY} - \Sigma_{YX} \Sigma_{XX} \Sigma_{XY}. \tag{7.55}$$

Gaussian distributions are

$$P(x) = \frac{1}{\sqrt{2\pi}} \exp\left\{ -\frac{(x-\mu)^2}{2\sigma^2} \right\}, \tag{7.56}$$

we generalize it as

$$P(x) = \frac{1}{Z(\mu, \sigma^2)} \exp(< t(\theta), \tau(x) >), \tag{7.57}$$

where $\tau(x) = <x, x^2>$, $t(\mu, \sigma^2) = <\frac{\mu}{\sigma^2}, -\frac{1}{2\sigma^2}>$,

$$Z(\mu, \sigma^2) = \sqrt{2\pi}\sigma \exp\left(\frac{\mu^2}{2\sigma^2} \right). \tag{7.58}$$

Linear exponential families is

$$P_\theta(x) = \frac{1}{Z(\theta)} \exp(< t(x), \theta >), \tag{7.59}$$

where

$$\Theta = \left\{ \theta \in \mathbf{R}^k, \int \exp(< t(x), \theta >) dx < \infty \right\}. \tag{7.60}$$

The exponential factor family is

$$\Phi_\theta(x) = A(x) \exp(< t(\theta), \tau(x) >), \tag{7.61}$$

and

$$P_\theta(x) \propto \prod_i \phi_{\theta_i}(x) = \prod_i A_i(x) \exp\left(\sum_i <t_i(\theta_i), \tau_i(x)>\right). \qquad (7.62)$$

Bayesian networks could be written in the corresponding way,

$$P(x|u) = \exp(t_{P(\mathbf{X}|\mathbf{U})}(\theta), \tau_{P(\mathbf{X}|\mathbf{U})}(x, u)). \qquad (7.63)$$

Entropy is

$$H(X) = \ln Z(\theta) - <E(\tau(X)), t(\theta)>. \qquad (7.64)$$

Relative entropy is

$$D(P_{\theta_1}||P_{\theta_2}) = E_{P(\theta_1)}\left[\ln\left(\frac{P_{\theta_1}(\mathcal{X})}{P_{\theta_2}(\mathcal{X})}\right)\right] = -\ln\frac{Z(\theta_1)}{Z(\theta_2)} + <E_{P(\theta_1)}(\tau(\mathcal{X})), t(\theta_1) - t(\theta_2)>. \qquad (7.65)$$

Information projection is

$$Q^I = \arg\min_{Q\in\mathcal{Q}} D(Q||P). \qquad (7.66)$$

Moment projection is

$$Q^M = \arg\min_{Q\in\mathcal{Q}} D(P||Q). \qquad (7.67)$$

If G_ϕ is an empty graph, $Q^M = \arg\max_{Q\in G_\phi} D(P|Q)$, then

$$Q^M(X_1, X_2, \ldots, X_n) = P(X_1)P(X_2)\ldots P(X_n). \qquad (7.68)$$

Therefore,

$$D(P||Q) = -H_P(\mathcal{X}) + E_P[\ln Q(\mathcal{X})] \geq D(P|Q^M). \qquad (7.69)$$

Furthermore, if and only if $Q_i(X) = P_i(X)$, then

$$D(P|Q) = D(P|Q^M), \qquad (7.70)$$

i.e., $Q = Q^M$, additionally,

$$D(P|Q_\theta) - D(P|Q_{\theta'}) = D(Q_{\theta'}||Q_\theta) \geq 0. \qquad (7.71)$$

7.5 Questions

Question 1. What are the aim and feature of Restricted Boltzmann machines (RBMs)? Why are deep Boltzmann machines thought as multilayer perceptron with restricted Boltzmann machines?

Question 2. What are the directed graphs and undirected graphs? Please give an example for each category.

Question 3. What is the difference between MRF and CRF?

Question 4. What are the linear exponential families? Please give an example.

Question 5. What are the relationships between relative entropy, information projection, and moment projection?

References

1. Ackley DH, Hinton GE, Sejnowski TJ (1987) A learning algorithm for Boltzmann machines. In: Readings in computer vision, pp 522–533
2. Fischer A, Igel C (2012) An introduction to restricted Boltzmann machines. In: Iberoamerican congress on pattern recognition, pp 14–36
3. Blake A, Rother C, Brown M, Perez P, Torr P (2004) Interactive image segmentation using an adaptive GMMRF model. In: European conference on computer vision. Springer, pp 428–441
4. Li S (2009) Markov random field modeling in image analysis. Springer, Berlin
5. Koller D, Friedman N (2009) Probabilistic graphical models. MIT Press, Cambridge
6. Hinton GE, Osindero S, Teh YW (2006) A fast learning algorithm for deep belief nets. Neural Comput 18(7):1527–1554
7. Sarikaya R, Hinton GE, Deoras A (2014) Application of deep belief networks for natural language understanding. IEEE/ACM Trans Audio Speech Lang Process 22(4):778–784
8. Goodfellow I, Bengio Y, Courville A (2016) Deep learning. MIT Press, Cambridge
9. Ertel W (2017) Introduction to artificial intelligence. Springer International Publishing, New York
10. Norvig P, Russell S (2016) Artificial intelligence: a modern approach, 3rd edn. Prentice Hall, Upper Saddle River
11. Chen LC, Papandreou G, Kokkinos I, Murphy K, Yuille AL (2018) DeepLab: semantic image segmentation with deep convolutional nets, atrous convolution, and fully connected CRFs. IEEE Trans Pattern Anal Mach Intell 40(4):834–848
12. Zheng S, Jayasumana S, Romera-Paredes B, Vineet V, Su Z, Du D, Torr PH (2015) Conditional random fields as recurrent neural networks. In: IEEE ICCV, pp 1529–1537

Transfer Learning and Ensemble Learning

8.1 Transfer Learning

8.1.1 Transfer Learning

In this chapter, we will introduce how to use well-trained parameters to test a new model. We hope transfer learning could save our computing time and costs. After transfer learning, the performance of new models may be good, or may not, thus we have to train the new models again and improve the model using new dataset.

Transfer learning is a new model of machine learning, which is used to change the training targets. We reuse our previously trained models and save our computing resources. In the past, a lot of work has been developed for deep learning. The best paper from CVPR'18 is about Taskonomy which is based on transfer learning. In transfer learning, the questions are: what is transfer learning? what will be transferred? when to be transferred? how to implement the transfer?

Transfer learning is a machine learning method where a model developed for a task is reused as the starting point for a model on a second task [1]. Transfer learning extracts knowledge (i.e., parameters, features, samples, instance, etc.) from a task and applies it to a new task. Transfer learning is famous for its unique features of a model and its parameters are availability. After a simple adjustment, the parameters could be confirmed and applied to new applications.

In transfer learning, corresponding to the discovered knowledge, we have sample transfer, instance transfer, parameter transfer, feature transfer, etc. According to the labels available and domain knowledge, we group transfer learning to many categories, e.g., supervised learning, unsupervised learning, reinforcement learning. Transfer AdBoost (TrAdBoost) is a typical example.

Transfer learning is subject to the labels of source task and target task. Transfer learning allows domain $\mathbf{D}$, task $\mathbf{T}$, and distribution used in training and test to be different. Because the domains are distinct, the tasks will be dissimilar.

Given domain $\mathbf{D} = \{\mathbf{X}, P(X), X \in \mathbf{X}\}$, $\mathbf{D}_S \neq \mathbf{D}_T$ implies $\mathbf{X}_S \neq \mathbf{X}_T$ or $P_S(X) \neq P_S(Y)$.

© The Author(s), under exclusive license to Springer Nature Switzerland AG 2021
W. Q. Yan, *Computational Methods for Deep Learning*, Texts in Computer Science,
https://doi.org/10.1007/978-3-030-61081-4_8

Given task $\mathbf{T} = \{\mathbf{Y}, P(Y|X), Y \in \mathbf{Y}\}$, $\mathbf{T}_S \neq \mathbf{T}_T$ implies $\mathbf{Y}_S \neq \mathbf{Y}_T$ or $P(Y_S|X_S) \neq P(Y_T|X_T)$.

If $\mathbf{D}_S = \mathbf{D}_T$, then $\mathbf{T}_S = \mathbf{T}_T$.

If $\mathbf{D}_S \neq \mathbf{D}_T$, then $\mathbf{T}_S \neq \mathbf{T}_T$ or $P(X_S) \neq P(X_T)$.

If $\mathbf{T}_S \neq \mathbf{T}_T$, then $\mathbf{Y}_S \neq \mathbf{Y}_T$ or $P(Y_S|X_S) \neq P(Y_T|X_T)$, $Y_S \in \mathbf{Y}_S$, $Y_T \in \mathbf{Y}_T$.

We have three kinds of transfer learning: inductive learning, transductive learning, and unsupervised learning (clustering), their domains, tasks, and algorithms are all different. Correspondingly, there are multiple methods for transfer learning: Sample-based transfer learning, feature-based transfer learning, parameters-based transfer learning, etc.

Inductive transfer learning [1] aims to improve the learning of the target predictive function $f_T(\cdot)$ in $\mathbf{D}_T$ using the knowledge in $\mathbf{D}_S$ and $\mathbf{T}_S$, where $\mathbf{T}_S \neq \mathbf{T}_T$. In inductive transfer learning, we transfer the samples, knowledge, and parameters.

Transductive learning [1] refers to the situation where all test data are required to be seen at training time, and the learned model cannot be reused for future data. Transductive learning could be applied to knowledge and parameter transfer.

Transductive transfer learning aims to improve the learning of the target predictive function $f_T(\cdot)$ in $\mathbf{D}_T$ using the knowledge in $\mathbf{D}_S$ and $\mathbf{T}_S$, where $\mathbf{D}_S \neq \mathbf{D}_T$ and $\mathbf{T}_S = \mathbf{T}_T$.

Unsupervised transfer learning (clustering or dimensionality reduction) [1] aims to improve the learning of the target predictive function $f_T(\cdot)$ in $\mathbf{D}_T$ using the knowledge in $\mathbf{D}_S$ and $\mathbf{T}_S$, where $\mathbf{T}_S \neq \mathbf{T}_T$ and $\mathbf{Y}_S$ and $\mathbf{Y}_T$ are not observable. No labelled data is observed in the source and target domains in training.

For transfer learning, a network is needed as well as training data and training algorithms. Training options should be specified when a network is being trained. Most of the time, training data is needed, which needs a huge workload to collect the training data and data augmentation. The training rate is related to how fast the network will be converged and what are the final trained parameters of the network.

In transfer learning, we need to estimate and evaluate the outcomes of transfer learning. We need to ensure that the results will be much better.

8.1.2 Taskonomy

The taskonomy paper (Disentangling Task Transfer Learning) was published in IEEE CVPR 2018 [2] which was awarded as the best paper award. In the work, taxonomy was created using a four steps process:

- **Step 1**. **Task Specific Modeling**: A task-specific network for each task is trained.
- **Step 2**. **Transfer Modeling**: All feasible transfers between sources and targets are trained.
- **Step 3**. **Ordinal Normalization**: The task affinities acquired from transfer function performances are normalized.

- **Step 4. Computing the Global Taxonomy**: A hypergraph is synthesized which can predict the performance of any transfer policy and optimize for the optimal one.

In taskonomy, the transfer operation is such a function that a small readout function $D_{s \to t}$ is trained to map representations of source task's frozen encoder to target task's labels.

Given a source task s and a target task t, where $s \in \mathbf{S}$ and $t \in \mathbf{T}$, a transfer network learns a small readout function for t given a statistic image computed for s.

$$D_{s \to t} = \arg \min_{\theta} E_{I \in D}(L_t(D_\theta(E_s(I)), f_t(I))), \tag{8.1}$$

where $f_t(I)$ is ground truth of t for image I. $D_{s \to t}$ is parameterized by $\theta_{s \to t}$ minimizing the loss $\mathbf{L}_t$.

8.2 Siamese Neural Networks

Different from transfer learning, Siamese networks or twin networks are typically used for the tasks that involve finding the relationship between two comparable and similar things, such as handwritten checks, face recognition, object tracking, and similar document matching, etc. Similarity measures are used based on a pair of twin networks. The objective of the Siamese network is to output a feature vector for each image so that the feature vectors are similar for similar images and different for dissimilar images. In this way, the network can discriminate the two input images.

Siamese networks are particularly useful in cases where there are large numbers of classes with small numbers of observations. Siamese networks can also be applied to dimensionality reduction

A Siamese network is a type of deep learning network which uses two or more identical subnetworks that have the same architecture and share the same parameters and weights. In that vein, two different input vectors will be computed for comparing output vectors, one of the output vectors is treated as the baseline, the other will be compared by computing the distance. The loss function is defined with squared Euclidean distance. The goal of Siamese network is to minimize a distance metric for similar objects and maximize for distinct ones.

$$L(\mathbf{x}_i, \mathbf{x}_j) = \begin{cases} \min(\| f(\mathbf{x}_i) - f(\mathbf{x}_j) \|_2) & \mathbf{x}_i = \mathbf{x}_j \\ \max(\| f(\mathbf{x}_i) - f(\mathbf{x}_j) \|_2) & \mathbf{x}_i \neq \mathbf{x}_j \end{cases} \tag{8.2}$$

For a half-twin network,

$$L(\mathbf{x}_i, \mathbf{x}_j) = \begin{cases} \min(\| f(\mathbf{x}_i) - g(\mathbf{x}_j) \|_2) & \mathbf{x}_i = \mathbf{x}_j \\ \max(\| f(\mathbf{x}_i) - g(\mathbf{x}_j) \|_2) & \mathbf{x}_i \neq \mathbf{x}_j \end{cases} \tag{8.3}$$

where i and j are indexes of two input vectors from the same dataset, $f(\cdot)$ and $g(\cdot)$ are the scoring functions implemented by using the twin network and half-twin network, respectively. More generally, the loss function is often approximated as a Mahalanobis distance for a linear space as

$$L(\mathbf{x}_i, \mathbf{x}_j) = (\mathbf{x}_i - \mathbf{x}_j)^\tau \cdot \mathbf{M} \cdot (\mathbf{x}_i - \mathbf{x}_j),\tag{8.4}$$

where $\mathbf{M}$ is a matrix from the Siamese network.

The dimension-reduced features allow the network to discriminate images that are similar and dissimilar. MATLAB provides a Siamese network for a demonstration to compare images and an example of dimensionality reduction.

The Siamese network reduces the dimension of input images and outputs the dimension-reduced images with the same label. The network is able to discriminate images that are similar and dissimilar. Because of two inputs and similarity measurement, siamese networks have been applied to visual object tracking in computer vision [3]. By measuring the similarity between the exemplar and each part of the search image, a map of similarity score can be generated by using the twin network.

Tracking objects are regarded as learning similarity problems. If the Siamese network has been chosen as the tracking network [3, 4], the algorithm is only working for single-object tracking, however, it can be combined with other deep learning algorithms such as fully connected neural network (FCNN), region proposal network (RPN), LSTM, etc. SiamRPN + LSTM algorithms have the highest MOTA (Multiple object tracking accuracy).

8.3   Ensemble Learning

One classifier is not enough in machine learning, because it only reflects one aspect and has mistakes or shortcomings. Multiple learners work together will make the classification better. The previous weaker classifiers, after the ensemble learning, will be much stronger. The classification of a classifier at least should be above 50%, the stronger classification should be beyond 50%. The essential ensemble learning methods include averaging, weighted average, and majority voting.

Averaging is a method based on a simple average over all the predictions from the different classifiers. Weighted average is based on the weights which are proportional to a classifier's capability and performance. In the majority voting method, the prediction is based on the most frequent class.

A set of diverse learners differ in their decisions so that they complement each other [5]. We combine multiple learners and free ourselves from taking a decision. (1) Draw random training sets from the given samples or training dataset (e.g., Bagging, etc.) (2) Train further base learners (e.g., boosting, cascading, etc.) (3) Mix multiple experts.

$$y = f(d_1, d_2, \ldots, d_L | \Phi)\tag{8.5}$$

$$c = \underset{i=1,2,\ldots,K}{\arg\max} \; y_i,\tag{8.6}$$

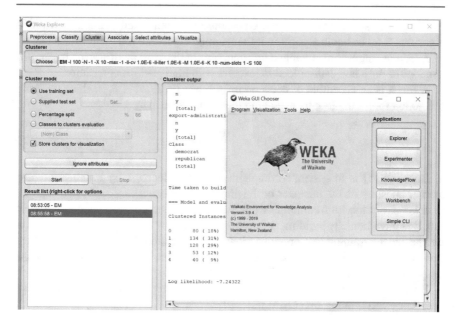

Fig. 8.1 An ensembling method provided by the famous machine learning software: Weka

where $f(\cdot)$ is the combining function with Φ denoting its parameters, c is the returned class number. Classifier combination rules include the operations $\sum(\cdot)$, $\max(\cdot)$, $\min(\cdot)$, $\prod(\cdot)$ and the simple voting $w_i = w_j \in \{1, 0\}$. If we use the combination $\sum(\cdot)$, mixing multiple learners is

$$y_i = \sum_j w_j d_{ji}, \sum_j w_j = 1, w_j \geq 0. \tag{8.7}$$

We have multiple methods to combine learners together. The application will decide which way of the combination will be applied.

Stacking refers to stack standardization. All classes have the same attributes. Stacking combines multiple models and multiple attributes together. Stacking mixes multiple models via a meta-model [5]. The meta-model is trained based on outputs of the base learners. Stacking achieves higher accuracy than using individual classifiers.

Stacking source code is available for public. Weka also has the corresponding function. Besides, the ensemble learning methods which Weka offers include Bagging, random forest, AdaBoost, voting. See the Weka interface with an ensembling method in Fig. 8.1.

Bootstrapping [5] is any test or metric that relies on random sampling with replacement. Bootstrapping performs inference about a sample from resampled data. Bootstrapping is a straightforward way to derive estimates of standard errors and confidence intervals.

Bagging (Bootstrap Aggregating) [5] is a voting method whereby base learners are made differently by training them over various training sets.

Bagging has each model in the ensemble vote with equal weight. A learning algorithm is an unstable algorithm if the learning algorithm has a high variance.

A learning algorithm is stable if different runs of the same algorithm on resampled versions of the same dataset lead to learners with high positive correlation.

A forest is an ensemble of decision trees $\mathbf{F} = \{T_1, T_2, \ldots, T_n\}$, which deliver a prediction for a sample x by averaging the output of each tree,

$$p_{\mathscr{F}}(y|x) = \frac{1}{k} \sum_{h=1}^{k} p_{T_h}(y|x). \tag{8.8}$$

Boosting [5] is to generate complementary base learners by training the next learner boosting on the mistakes of the previous learners.

A boosting algorithm combines weak learners to generate strong learner. Boosting involves incrementally building an ensemble learning by training each new model to emphasize the training that previous models misclassified. Reducing misclassification is an effective way of boosting classification. In the boosting classification, the weights of classifications are different.

Boosting is interpreted as an optimization algorithm based on a suitable cost function. For a training set $\{(x_i, y_i)\}, i = 1, \ldots, n,$

$$F(x) \overset{\Delta}{=} \sum_{i=1}^{M} \gamma_i \cdot h_i, \tag{8.9}$$

where h_i is the base learner.

$$\hat{F}(x) = \arg\max_{F} \mathbf{E}_{x,y}[L(y, F(x))], \tag{8.10}$$

where $L(\cdot)$ is the cost function.

Using the steepest descent $\nabla_{F_{m-1}} L(\cdot), m = 1, 2, \ldots$, we have

$$F_m(x) = F_{m-1}(x) - \gamma_m \cdot \sum_{i=1}^{n} \nabla_{F_{m-1}} L(y_i, F_{m-1}(x_i)), \tag{8.11}$$

where

$$\gamma_m = \arg\max_{\gamma} \sum_{i=1}^{n} L(y_i, F_{m-1}(x_i)) - \gamma \nabla_{F_{m-1}} L(y_i, F_{m-1}(x_i))). \tag{8.12}$$

Thus, derivatives or gradients need to be calculated. This algorithm could be written in computable source code.

AdaBoost (Adaptive Boosting) takes use of the same training set over and over, the classifiers should be simplified so that they do not overfit. The dataset should not be changed. AdaBoost can combine an arbitrary number of base learners based on weights. The success of AdaBoost is due to its salient property of increasing margin.

Cascading is a multistage method where there is a sequence of classifiers and the next one is used only when the preceding ones are not confident [5]. Cascading has been applied to face detection, this ensemble learning could be applied to general visual object detection. Cascading generates a rule (or rules) to explain a large part of the instances as cheaply as possible and stores the rest as exceptions.

For ensemble learning, all open Python source codes are available from the scikit-learn.org website at: https://scikit-learn.org/stable/modules/classes.html♯ module-sklearn.ensemble. See the screenshot of the scikit emsembling methods in Fig. 8.2.

Scikit-learn is a free software machine learning library for the Python programming language. The library was developed based on NumPy, SciPy, and matplotlib.

MATLAB can meld results from many weak learners into one high-quality ensemble predictor by using ensemble learning. The methods include bootstrap aggregation (Bagging), random forest, boosting algorithms, etc. Boosting algorithms encapsulate adaptive boosting, gentle adaptive boosting, adaptive logistic regression, linear programming boosting, least squares boosting, robust boosting, random undersampling boosting, etc. See the screenshot of the MATLAB emsembling methods based on regression tree in Fig. 8.3.

8.4 Important Work in Deep Learning

In this section, we will detail the work published in Nature and Science. Publications on the journal Nature include:

- B. Zhu, et al. (2018) Image reconstruction by domain-transform manifold learning. Nature, 555: 487–492
- S. Webb. (2018) Deep learning for biology. Nature, 554: 555–557
- Y. LeCun, Y. Bengio, G. Hinton. (2015) Deep learning. Nature, 521: 436–444
- M. Littman (2015) Reinforcement learning improves behavior from evaluative feedback, Nature, 521: 445–451
- V. Mnih, et al. (2015) Human-level control through deep reinforcement learning, Nature, 518: 529–533
- D. Rumelhart, G. Hinton, et al. (1986) Learning representations by back-propagating errors. Nature, 323: 533–536

The works published on the journal Science include:

- D. George, et al. (2017) A generative vision model that trains with high data efficiency and breaks text-based CAPTCHAs. Science, 358 (6368)
- M. I. Jordan and T. M. Mitchell. (2015) Machine learning: Trends, perspectives, and prospects. Science, 349 (6245): 255–260
- G. Hinton, R. Salakhutdinov. (2006) Reducing the dimensionality of data with neural networks. Science, 313(5786):504–507

Deep learning was thought that it could be applied to computer vision, natural language processing, robot control, and other applications [6]. The supervised learning methods include decision trees/forests, logistic regression, deep neural networks,

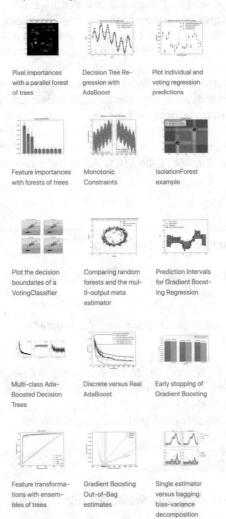

Ensemble methods

Examples concerning the `sklearn.ensemble` module.

Pixel importances with a parallel forest of trees

Decision Tree Regression with AdaBoost

Plot individual and voting regression predictions

Feature importances with forests of trees

Monotonic Constraints

IsolationForest example

Plot the decision boundaries of a VotingClassifier

Comparing random forests and the multi-output meta estimator

Prediction Intervals for Gradient Boosting Regression

Multi-class Ada-Boosted Decision Trees

Discrete versus Real AdaBoost

Early stopping of Gradient Boosting

Feature transformations with ensembles of trees

Gradient Boosting Out-of-Bag estimates

Single estimator versus bagging: bias-variance decomposition

Plot the decision surfaces of ensembles of trees on the iris dataset

Combine predictors using stacking

Fig. 8.2 Ensembling methods provided by scikit-learn

Regression Tree Ensembles — Examples

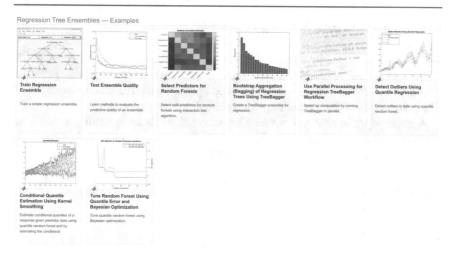

Fig. 8.3 Ensembling methods provided by MATLAB

Bayesian classifiers, Boosting, etc. Deep neural networks are multilayer networks with threshold units. Deep learning makes use of gradient-based optimization algorithms to adjust parameters throughout a multilayered network based on errors at its output. The internal layers of deep networks can be viewed as providing learned representations of the input data.

Unsupervised learning is mainly applied to solve the problems such as clustering and dimensionality reduction, these mathematical approaches include PCA (a linear dimension reduction method), manifold learning (a nonlinear dimension reduction method), autoencoders, etc.

Reinforcement learning is one of three basic machine learning paradigms, alongside supervised learning and unsupervised learning. In reinforcement learning, the information which is available from training data is intermediate between supervised and unsupervised learning. The training data in reinforcement learning are assumed to provide only an indication whether an action is correct or not; if an action is incorrect, there remains the problem of finding the correct action.

In the setting of sequences of inputs, it is assumed that reward signals refer to the entire sequence. Reinforcement learning generally makes use of ideas that are familiar from the control theory, such as policy iteration, value iteration, rollouts, and variance reduction, with innovations arising to address the specific needs of machine learning.

A recommendation system is based on data that indicates the links between a set of users (e.g., people) and a set of items (e.g., products). The machine learning problem is to suggest other items to a given user based on the data across all users.

8.5   Awarded Work in Deep Learning

In this section, we mainly emphasize the work on IEEE CVPR and IEEE ICCV conferences. IEEE ICCV has the best paper award: Marr award. The Marr prize is a biennial conference award in computer vision given by the IEEE conference ICCV. The prize is one of the top honours for a computer vision researcher.

A good paper is reflected in many aspects such as its idea, writing, references, equations, tables, and figures, etc. The key is how the paper attracts readers and what the impact could be generated from this paper.

- Disentangling Task Transfer Learning, IEEE CVPR 2018 [2]
- SimGAN [7]
- Mask R-CNN [8]
- DenseNets [9].

8.6   Questions

Question 1. What's knowledge discovery? How to use the learned knowledge in transfer learning?

Question 2. Based on labels, how to group transfer learning models?

Question 3. How to train multiple multi-label classification?

Question 4. What are the differences between DenseNets and ResNets?

Question 5. What are the applications of deep neural networks in ensemble learning?

References

1. Pan S, Yang Q (2010) A survey on transfer learning. IEEE Trans Knowl Data Eng 22(10):1345–1359
2. Zamir A et al. Taskonomy: disentangling task transfer learning. In: CVPR'18
3. An N (2020) Anomalies detection and tracking using Siamese neural networks. Master thesis, Auckland University of Technology, New Zealand
4. Valmadre J, Bertinetto L, Henriques J, Vedaldi A, Torr P (2017) End-to-end representation learning for correlation filter based tracking. In: IEEE conference on computer vision and pattern recognition (CVPR), pp 2805–2813
5. Alpaydin E (2009) Introduction to machine learning. MIT Press, Cambridge

6. Jordan MI, Mitchell TM (2015) Machine learning: trends, perspectives, and prospects. Science 349(6245):255–260
7. Shrivastava A et al. Learning from simulated and unsupervised images through adversarial training. In: CVPR'17
8. Takeda F, Omatu S (1995) A neuro-paper currency recognition method using optimized masks by genetic algorithm. In: IEEE international conference on systems, man and cybernetics, vol 5, pp 4367–4371
9. Huang G, Liu Z, Weinberger KQ, van der Maaten L (2017) Densely connected convolutional networks. In: IEEE CVPR, vol 1(2), pp 3

Glossary

Activation function In artificial neural networks, the activation function of a node defines the output of that node given an input or set of inputs.

AdaBoost Adaptive Boosting, a voting method for training a boosted classifier

Autoencoder An autoencoder is a neural network that learns to copy its input to its output.

Average pooling Calculating the average for each patch of the feature map.

Atlas A specific collection of charts which covers a manifold

Bagging Bootstrap AGGregatING, a machine learning ensemble meta-algorithm designed to improve the stability and accuracy of machine learning algorithms used in statistical classification and regression

Banach spaces Complete normed vector spaces

Bayesian inference A method of statistical inference in which Bayes' theorem is used to update the probability for a hypothesis as more evidence or information becomes available

Bayesian learning Using Bayes' theorem to determine the conditional probability of a hypothesis given evidence or observations

Bayesian network A decision network, which is a type of statistical model that represents a set of variables and their conditional dependencies via a directed acyclic graph (DAG)

Boltzmann machine Stochastic Hopfield network with hidden units, a type of stochastic recurrent neural network

Boltzmann distribution The distribution maximizes the entropy.

Boosting An ensemble meta-algorithm for primarily reducing bias, and also variance [1] in supervised learning, and a family of machine learning algorithms that convert weak learners to strong ones

Bootstrapping Test or metric that uses random sampling with replacement

CapsNet Capsule neural network, which is a type of artificial neural network (ANN) for better modelling hierarchical relationships of an object

Capsule A set of neurons that are individually activated for various properties of a type of objects

© The Editor(s) (if applicable) and The Author(s), under exclusive license
to Springer Nature Switzerland AG 2021
W. Q. Yan, *Computational Methods for Deep Learning*, Texts in Computer Science,
https://doi.org/10.1007/978-3-030-61081-4

Cascading   A particular case of ensemble learning based on the concatenation of several classifiers, using all information collected from the output from a given classifier as additional information for the next classifier

Chart   An invertible map between a subset of the manifold and a simple space such that both the map and its inverse preserve the desired structure

Clique tree   The junction tree algorithm, a method in machine learning to extract marginalization from general graphs

CNN   Convolutional neural network

Convex   The line segment between any two points on the graph of the function lies one side of the graph between the two points.

ConvNet   Convolutional neural network

Convolution   A mathematical operation for two functions produces a third function expressing how the shape of one is modified by the other.

DAG   Directed acyclic graph, a finite directed graph with no directed cycles

DBM   Deep Boltzmann machine, a type of binary pairwise Markov random field which is an undirected probabilistic graphical model with multiple layers of hidden random variables

Decision tree   A decision support tool that uses a tree-like model of decisions and their possible consequences, including chance event outcomes, resource costs, and utility

Decision rule   A function which maps an observation to an appropriate action.

Deep learning   Deep neural network has a powerful ability of nonlinear processing using a cascade of multiple layers network for feature transformation and end-to-end learning.

DRL   Deep reinforcement learning, using deep learning and reinforcement learning principles to create efficient algorithms

Double Q-learning   An off-policy reinforcement learning algorithm, where a different policy is used for value evaluation than what is used to select the next action

Dynamic Bayesian network   A Bayesian network represents sequences of variables.

EKF   Extended Kalman filter, nonlinear Kalman filter which linearizes about an estimate of the current mean and covariance

Ensemble learning   The process by which multiple models are strategically generated and combined to solve a particular computational intelligence problem

Entropy   A measure of the unpredictability of the state, or equivalently, of its average information content.

Event   In textual topic detection and extraction, an event is something that happened somewhere at a certain time.

Exponential family   A set of distributions, where the specific distribution varies with the parameter

Factorization   A product of factors over cliques in the graph

Fourier transform   A mathematical transform that decomposes a function into its constituent frequencies

Fuzzy optimization   A mathematical model which deals with transitional uncertainty and information deficiency uncertainty

GCD   The greatest common divisor of two or more integers, which are not all zero, is the largest positive integer that divides each of the integers.

Genetic algorithm   A metaheuristic inspired by the process of natural selection that belongs to the larger class of evolutionary algorithms.

Gibbs distribution   A probability distribution or probability measure

Gibbs measure   The unique statistical distribution that maximizes the entropy for a fixed expectation value of the energy

Global optimization   The task of finding the absolutely best set of admissible conditions to achieve your objective, formulated in mathematical terms

Hausdorff space   A topological space in which each pair of distinct points is separated by a disjoint open set

Hilbert spaces   An inner product space which is complete as a metric space

Induced subgraph $G[S]$ is the graph whose vertex set is S and whose edge set consists of all of the edges in E that have both endpoints in $S \subset G = (V, E)$.

Influence diagram   A compact graphical and mathematical representation of a decision situation

Isometry   congruence, a distance-preserving transformation between metric spaces, usually assumed to be bijective

Joint entropy   A measure of the uncertainty associated with a set of variables

Kalman filter   The optimal linear estimator for linear system models with additive independent white noise in both the transition and the measurement systems

KL divergence   Kullback–Leibler divergence or relative entropy, a measure of how one probability distribution is different from a second and reference probability distribution

Latent variables   The variables that are not directly observed but are rather inferred (through a mathematical model) from other variables that are observed (directly measured)

LCM   The smallest common multiple of two integers a and b is the smallest positive integer that is divisible by both a and b.

Lipschitz continuity   A strong form of uniform continuity for functions

Linear Dynamical System   A dynamic Bayesian network where all of the dependencies are linear Gaussian

Linear programming   A technique for the optimization of a linear objective function, subject to linear equality and linear inequality constraints

LSTM   Long short-term memory network

MAP   Maximum a posteriori probability estimate, an estimate of an unknown quantity, that equals the mode of the posterior distribution

Manifold   A topological space with the property that each point has a neighbourhood, a second countable Hausdorff space that is locally homeomorphic to Euclidean space.

Manifold learning   An approach for nonlinear dimensionality reduction

Markov chain   A stochastic model describing a sequence of possible events in which the probability of each event depends only on the state attained in the previous event

Markov process   A stochastic process that satisfies the Markov property

Max pooling   Applying a max filter to non-overlapping subregions of the initial representation

MDP   Markov decision process, a discrete-time stochastic control process

Metadata   Data about data, namely additional information of a given set of data

Metric   A function that defines a concept of distance between any two members of the set, which are usually called points

Metric space   A set together with a metric on the set

MGU   Minimal Gated Unit

MLE   Maximum likelihood estimation, a method of estimating the parameters of a probability distribution by maximizing a likelihood function, under the assumed statistical model, the observed data is most probable.

MNIST   The Modified National Institute of Standards and Technology (NIST) dataset

MRF   Markov random field which is a set of random variables having a Markov property described by an undirected graph.

Multiobjective programming   A part of mathematical programming dealing with decision problems characterized by using multiple and conflicting objective functions that are to be optimized over a feasible set of decisions

Mutual information   A measure of the mutual dependence between the two variables

Naive Bayes Model   A family of simple probabilistic classifiers based on applying Bayes' theorem with strong independence assumptions

NLAR   Nonlinear autoregressive model

Norm   A real-valued function defined on the vector space

Normed space   A vector space over the real or complex numbers, on which a norm is defined.

Observable variable   Manifest variable in statistics, a variable that can be observed and directly measured

Orbifold   A generalization of manifold allowing for "singularities" in the topology

Padding   The filled region of an image boundary is applied to fill up the edge region with zero.

Parameter estimation   The process of using sample data to estimate the parameters of the selected distribution

Particle swarm optimization   A computational method that optimizes a problem by iteratively improving a candidate solution with regard to a given measure of quality with a population of candidates, moving the particles in the search space according to the position and velocity

Partition functions   A generating function for expectation values of various functions of the random variables

PGM Probabilistic graphical model or structured probabilistic model which is a model for a graph to expresses the conditional dependence structure between random variables

Q-learning A model-free reinforcement learning algorithm to learn a policy telling an agent what action to take under what circumstances

Random forests An ensemble learning method for classification and regression by constructing a multitude of decision trees at training time and outputting the class that is the mode of classification regression of the individual trees.

RBM Restricted Boltzmann machine, a variant of Boltzmann machine with the restriction that their neurons must form a bipartite graph

Regularization The process of adding information in order to solve an ill-posed problem or to prevent overfitting

Reinforcement learning Approximate dynamic programming or neuro-dynamic programming, which is teaching a software agent how to behave in an environment by telling it how good it's doing

ResNet Deep residual network

Reward function A function defines the goal for an agent.

RNN Recurrent neural network

Roots of a polynomial Those values of the variable that cause the polynomial to evaluate to zero

SARSA State–action–reward–state–action, an algorithm for learning a Markov decision process policy, used in the reinforcement learning area of machine learning

Siamese neural network Twin neural network, an artificial neural network that uses the same weights while working in tandem on two different input vectors to compute comparable output vectors

Single shot The tasks of visual object localization and classification are done in a single forward pass of the network.

Squashing function A function that squashes the input to one of the ends of a small interval

SSD Single shot multibox detector

Stride The step length of convolution operations

Target variable The variable whose values are to be modelled and predicted by other variables

Temporal difference learning A class of model-free reinforcement learning methods which learn by bootstrapping from the current estimate of the value function

Tensor A generalization of vectors and matrices to potentially higher dimensions

TensorFlow A framework to define and run computations involving tensors, represent tensors as n-dimensional arrays of base datatypes.

Time series analysis Analysing time series data in order to extract meaningful statistics and other characteristics of the data

Time series forecasting The use of a model to predict future values based on previously observed values

Transfer function This function is used for transformation purposes, from input nodes to the output of a neuron.

Transfer learning   A machine learning method where a model is developed for a task which is reused as the starting point for a model on a second task.

YOLO   You only look once

Index

Printed in the United States
by Baker & Taylor Publisher Services